Mathematics and Control Engineering of Grinding Technology

Mathematics and Its Applications (*East European Series*)

Mathematics and Control Engineering of Grinding Technology

Ball Mill Grinding

L. Keviczky

M. Hilger

and

J. Kolostori

Computer and Automation Institute, Hungarian Academy of Sciences, Budapest, Hungary

KLUWER ACADEMIC PUBLISHERS

DORDRECHT / BOSTON / LONDON

Library of Congress Cataloging in Publication Data

Keviczky, László.
Mathematics and control engineering of grinding technology : ball mill grinding / L. Keviczky, M. Hilger, and L. [i.e. J.] Kolostori.
p. cm. -- (Mathematics and its applications. East European series)
Bibliography: p.
Includes index.
ISBN 0-7923-0051-3
1. Size reduction of materials. 2. Ball mills. I. Hilger, Miklós. II. Kolostori, János. III. Title. IV. Series: Mathematics and its applications (D. Reidel Publishing Company). East European series.
TP156.S5K48 1989
621.9'2--dc19 88-30446

ISBN 0-7923-0051-3

Published by Kluwer Academic Publishers,
P.O. Box 17, 3300 AA Dordrecht, The Netherlands.

Kluwer Academic Publishers incorporates
the publishing programmes of
D. Reidel, Martinus Nijhoff, Dr W. Junk and MTP Press.

Sold and distributed in the U.S.A. and Canada
by Kluwer Academic Publishers,
101 Philip Drive, Norwell, MA 02061, U.S.A.

In all other countries, sold and distributed
by Kluwer Academic Publishers Group,
P.O. Box 322, 3300 AH Dordrecht, The Netherlands.

printed on acid free paper

Printed in The Netherlands

TO CSILLA

SERIES EDITOR'S PREFACE

'Et moi, ..., si j'avait su comment en revenir, je n'y serais point allé.'

Jules Verne

The series is divergent; therefore we may be able to do something with it.

O. Heaviside

One service mathematics has rendered the human race. It has put common sense back where it belongs, on the topmost shelf next to the dusty canister labelled 'discarded non-sense'.

Eric T. Bell

Mathematics is a tool for thought. A highly necessary tool in a world where both feedback and non-linearities abound. Similarly, all kinds of parts of mathematics serve as tools for other parts and for other sciences.

Applying a simple rewriting rule to the quote on the right above one finds such statements as: 'One service topology has rendered mathematical physics ...'; 'One service logic has rendered computer science ...'; 'One service category theory has rendered mathematics ...'. All arguably true. And all statements obtainable this way form part of the raison d'être of this series.

This series, *Mathematics and Its Applications*, started in 1977. Now that over one hundred volumes have appeared it seems opportune to reexamine its scope. At the time I wrote

> "Growing specialization and diversification have brought a host of monographs and textbooks on increasingly specialized topics. However, the 'tree' of knowledge of mathematics and related fields does not grow only by putting forth new branches. It also happens, quite often in fact, that branches which were thought to be completely disparate are suddenly seen to be related. Further, the kind and level of sophistication of mathematics applied in various sciences has changed drastically in recent years: measure theory is used (non-trivially) in regional and theoretical economics; algebraic geometry interacts with physics; the Minkowsky lemma, coding theory and the structure of water meet one another in packing and covering theory; quantum fields, crystal defects and mathematical programming profit from homotopy theory; Lie algebras are relevant to filtering; and prediction and electrical engineering can use Stein spaces. And in addition to this there are such new emerging subdisciplines as 'experimental mathematics', 'CFD', 'completely integrable systems', 'chaos, synergetics and large-scale order', which are almost impossible to fit into the existing classification schemes. They draw upon widely different sections of mathematics."

By and large, all this still applies today. It is still true that at first sight mathematics seems rather fragmented and that to find, see, and exploit the deeper underlying interrelations more effort is needed and so are books that can help mathematicians and scientists do so. Accordingly MIA will continue to try to make such books available.

If anything, the description I gave in 1977 is now an understatement. To the examples of interaction areas one should add string theory where Riemann surfaces, algebraic geometry, modular functions, knots, quantum field theory, Kac-Moody algebras, monstrous moonshine (and more) all come together. And to the examples of things which can be usefully applied let me add the topic 'finite geometry'; a combination of words which sounds like it might not even exist, let alone be applicable. And yet it is being applied: to statistics via designs, to radar/sonar detection arrays (via finite projective planes), and to bus connections of VLSI chips (via difference sets). There seems to be no part of (so-called pure) mathematics that is not in immediate danger of being applied. And, accordingly, the applied mathematician needs to be aware of much more. Besides analysis and numerics, the traditional workhorses, he may need all kinds of combinatorics, algebra, probability, and so on.

In addition, the applied scientist needs to cope increasingly with the nonlinear world and the extra mathematical sophistication that this requires. For that is where the rewards are. Linear models are honest and a bit sad and depressing: proportional efforts and results. It is in the non-linear world that infinitesimal inputs may result in macroscopic outputs (or vice versa). To appreciate what I am hinting at: if electronics were linear we would have no fun with transistors and computers; we would have no TV; in fact you would not be reading these lines.

There is also no safety in ignoring such outlandish things as nonstandard analysis, superspace and anticommuting integration, *p*-adic and ultrametric space. All three have applications in both electrical engineering and physics. Once, complex numbers were equally outlandish, but they frequently proved the shortest path between 'real' results. Similarly, the first two topics named have already provided a number of 'wormhole' paths. There is no telling where all this is leading - fortunately.

Thus the original scope of the series, which for various (sound) reasons now comprisese five subseries: white (Japan), yellow (China), red (USSR), blue (Eastern Europe), and green (everything else), still applies. It has been enlarged a bit to include books treating of the tools from one subdiscipline which are used in others. Thus the series still aims at books dealing with:

- a central concept which plays an important role in several different mathematical and/or scientific specialization areas;
- new applications of the results and ideas from one area of scientific endeavour into another;
- influences which the results, problems and concepts of one field of enquiry have, and have had, on the development of another.

As regards the present volume one can say that the raw materials are all at hand: empirical knowledge and experience, theory in the form of models and of various bits and pieces of mathematics, processing tools in the form of computers. This is the situation in many industrial and engineering fields. It remains to bring the three together and reap the rewards. That is precisely what the authors of the present volume have done for the topic of ball-mill grinding. And thus they show, and show well, that indeed there is an enormous range of mathematical tools ready (and eager) to be applied.

Perusing the present volume is not guaranteed to turn you into an instant expert, but it will help, though perhaps only in the sense of the last quote on the right below.

The shortest path between two truths in the real domain passes through the complex domain.

J. Hadamard

La physique ne nous donne pas seulement l'occasion de résoudre des problèmes ... elle nous fait pressentir la solution.

H. Poincaré

Never lend books, for no one ever returns them; the only books I have in my library are books that other folk have lent me.

Anatole France

The function of an expert is not to be more right than other people, but to be wrong for more sophisticated reasons.

David Butler

Bussum, October 1988 Michiel Hazewinkel

CONTENTS

1. INTRODUCTION

The plant measurements and the technological tests performed by the authors mainly in the cement industry, and their control engineering, simulation and modelling researches as well as industrial experience obtained with real control systems *made it possible*; the continuing deterioration of the available raw materials and the permanently increasing quality requirements of the products *made it necessary* to formulate a so-called control engineering theory of ball mill grinding by integrating the knowledge accumulated until now, and by combining algorithmizable and empirical theories.

Within this framework, on the basis of the available measurements, designed experiments and the relevant technical literature, we have tried to clarify which of the very sophisticated and serious problems of the theory of grinding belong to the field of empiricism and which are those where the theoretical background can be derived or had already been known earlier.

All this had to be achieved by forming a relationship between the classical and advanced theories of grinding on the one hand, and the technological and theoretical control engineering system approaches on the other.

The fundamentals of the technology (e.g. particle size distribution, specific surface, etc.) are treated only as far as they are needed for the understanding of mathematical models and control systems introduced later for users not expert in grinding engineering. These fundamentals are based mainly on the work and scientific results of Professor **BEKE**. A more detailed acquaintance with this discipline can be obtained from the numerous references given in the bibliography.

After the introduction and the derivation of the state equations of the batch grinding processes, the material flow models of open- and closed-circuit (ball) grinding mills are discussed, where - besides several references - we rely on the results of our research, simulation investigations and papers. These are valid mainly for the macrostructural models.

In the chapter dealing with the mathematical model of the classifier, the relevant results from the literature are summarized.

In these chapters special attention has been drawn to the results directly applicable in the - everyday and not specially cement industry - practice, and to proofs of the relationships that have been considered empirical until now. An example of the foregoing is, at batch grinding, how the economically optimal grinding time can be determined from the variation of particle size distribution. With knowledge of the grinding time, the chosen control variables (e.g. mill speed, ball diameter, distribution, etc.) can be optimized to maximize the quantity of the final product. An optimal product size distribution can be prescribed in advance which should be reached as closely as possible by the end of the batch process. This task can be combined with the simultaneous minimization of the desired grinding time, etc. With regard to the dynamic modelling of the chemical composition of ground materials, the effect of changes in the composition matrix is especially emphasized.

Models for investigating the average oxide compositions of products are given for modelling batch and continuous operation silos.These units are joint parts of grinding technologies very frequently in cement plants.

In the chapter discussing the advanced control systems of grinding equipments, the most typical solutions widely used in the practice are presented on the one hand, and, on the other hand, also emphasized are the application possibilities of the latest methods of advanced control theory (these are the adaptive optimal control algorithms) which can be applied even today by the help of advanced microcomputer technology. In this chapter the problems of composition, quantity control and fineness control are investigated.

The detailed derivations of the relationships are given in the Appendices for easier reading.

Computer control and optimization of technological processes involving grinding technology are very important and urgent but very complex tasks. The interdisciplinary description of grinding processes and the integrated knowledge introduced by this book can be considered to be a suitable theoretical basis to establish an *expert system* specialized for grinding.

SYMBOLS USED IN THE FIGURES:

NAME	BLOCK-DIAGRAM ELEMENTS	FUNCTION
INTEGRATOR	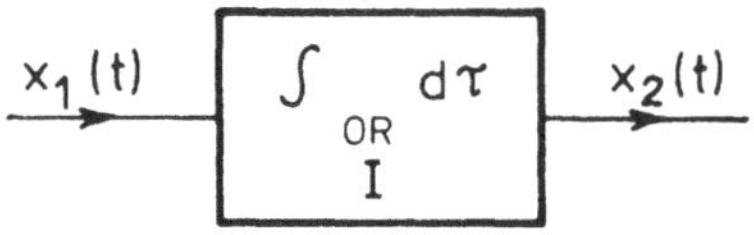 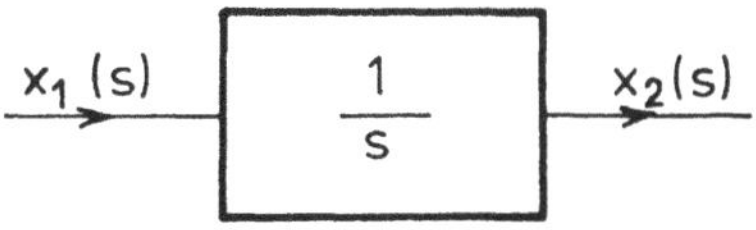	$x_2(t) = \int_0^t x_1(t)dt + x_2(0)$ $x_1(s) = \frac{x_2(s)}{s}$
MULTIPLICATOR	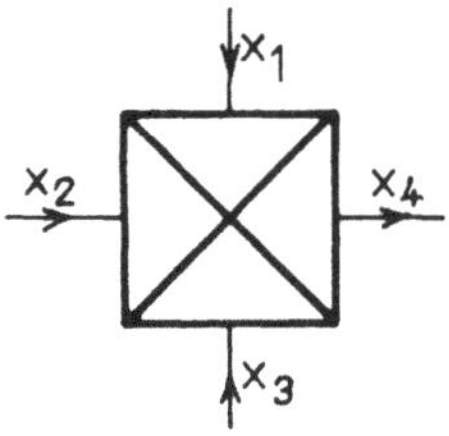	$x_4 = x_1 x_2 x_3$
DIVISOR	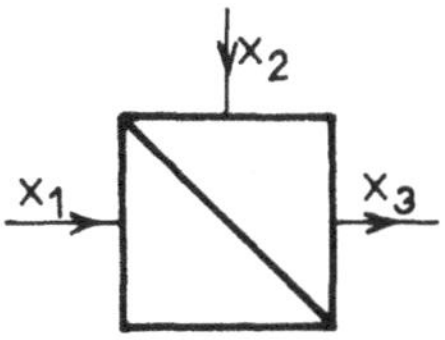	$x_3 = \frac{x_1}{x_2}$
TIME DELAY	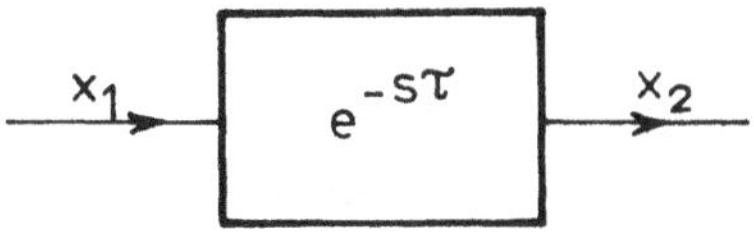	$x_2 = x_1(t-\tau)$
LIMIT	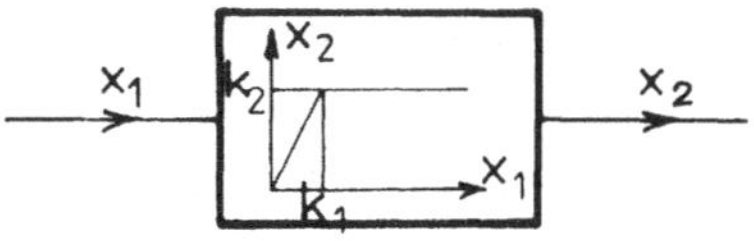	$x_2 = k_2$ IF $x_1 \geqslant k_1$

NAME	BLOCK-DIAGRAM ELEMENTS	FUNCTION
SWITCH	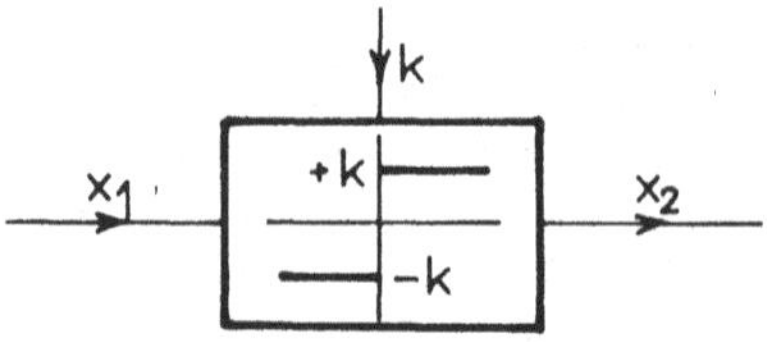	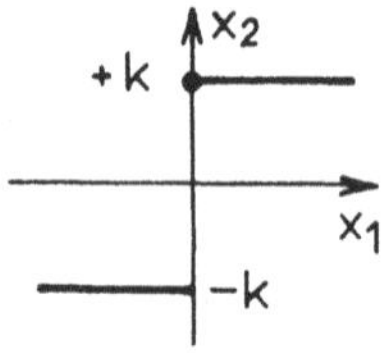
SIGNAL BRANCH	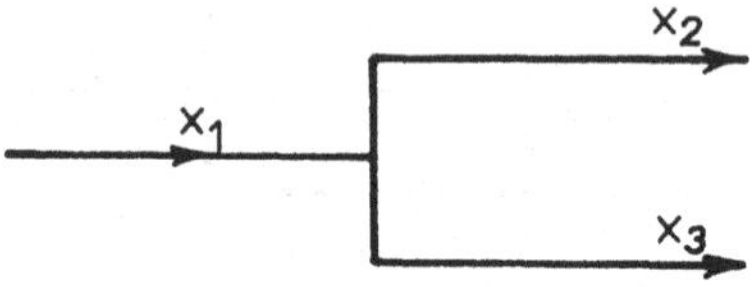	$x_1 = x_2 = x_3$
SUMMATOR	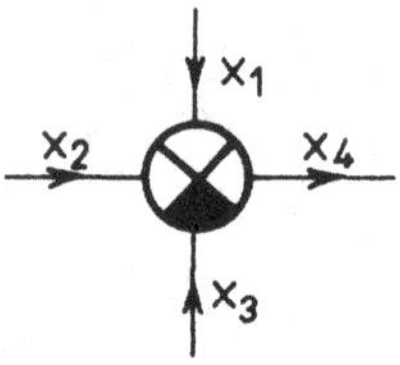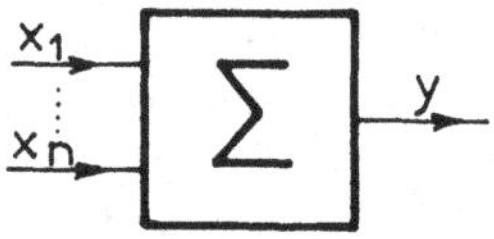	$x_4 = x_1 + x_2 - x_3$ $y = \sum_{i=1}^{n} x_i$
INPUT/OUTPUT SIGNALS OF AN ELEMENT		
FEEDBACK	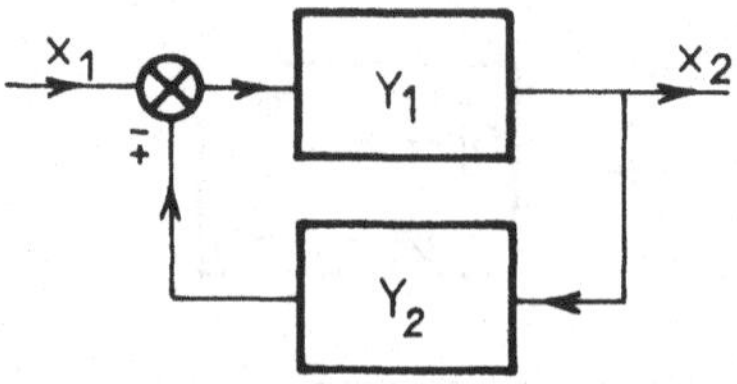	$x_2 = \frac{Y_1}{1 \pm Y_1 Y_2} x_1$

NAME	BLOCK-DIAGRAM ELEMENTS	FUNCTION
NONLINEARITY		$X_2 = N_i(X_1)$ where N_i is a nonlinear function
MULTIPLE INPUT MULTIPLE OUTPUT ELEMENT		
SUMMATOR	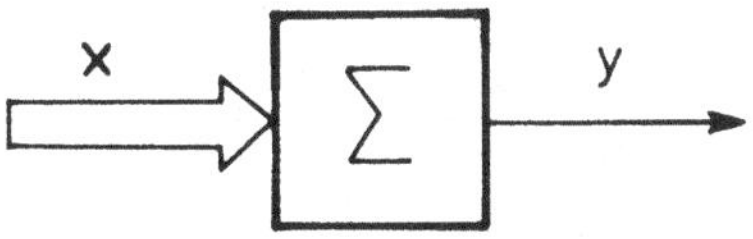	$y = \sum_{i=1}^{n} x_i$

2. PARTICLE SIZE DISTRIBUTION OF GROUND MATERIAL

The particle size composition of the crushed material can be given by a continuous function where the particle size is regarded as a probability variable. The size distribution can be represented by the passing (D) and retaining (R) functions (see Fig. 2-1). The ordinates give the weight in percent. of the material which is passed and retained by the relevant sieve with aperture d.

The derivative of the passing function

$$D'(d) = \frac{\partial D(d)}{\partial d} \qquad (2.1)$$

gives the probability of the particle size d. The passing function D is generally called the cumulative or (simply) distribution function based on the definition of the distribution function well-known from statistics, since at a given d it gives the probability of occurrence of particles smaller than d (i.e. the proportion of particles smaller than a certain size). On the basis of the above analogy function D' can be called the density function (see (2.1)).

A very large number of investigations showed that the particle size distribution can be well approximated by functions having two constants. Numerous continuous functions have been postulated to describe the size distribution of products of comminution.

In the Anglo-Saxon practice the formula of **GATES-GAUDIN- SCHUMANN (GGS)** [67],[169],[170]

$$D(d) = \left(\frac{d}{d_o}\right)^m \qquad (2.2)$$

or **GAUDIN-MELOY** [68]

$$D(d) = 1 - \left(1 - \frac{d}{d_o}\right)^m \qquad (2.3)$$

is used. (Here and in the following, d_o means the maximum particle size (or the size of the original specimen)).

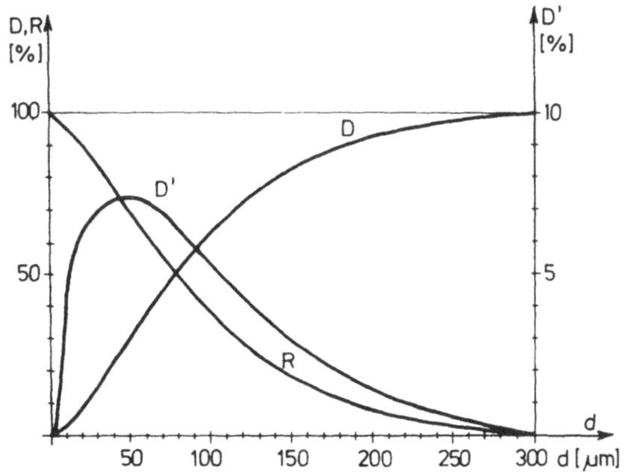

Fig. 2-1 Characteristic functions of size distribution

In Europe the formula of **ROSIN-RAMMLER (RR)** [187]

$$D(d) = 1 - \exp(-b\, d^m) \qquad (2.4)$$

or its other version derived by **BENNETT** [141]

$$D(d) = 1 - \exp\left[-\left(\frac{d}{d_o}\right)^m\right] \qquad (2.5)$$

is mainly used. The latter can be easily normalized by ensuring that $D(d_o)=1$, which leads to the form elaborated by **BROADBENT-CALCOTT** [142]

$$D(d) = \frac{e}{e-1}\left\{1 - \exp\left[-\left(\frac{d}{d_o}\right)^m\right]\right\} \quad . \qquad (2.6)$$

The function of **HARRIS** [58,80] has also to be mentioned

$$D(d) = 1 - \left[1 - \left(\frac{d}{d_o}\right)^m\right]^r \qquad (2.7)$$

as well as the log-normal **(LN)** distribution derived by **KOLMOGOROV** in a theoretical way under simplified conditions

$$D(d) = \varphi\left[\ln\left(\frac{d}{d_o}\right)^m\right] \quad . \qquad (2.8)$$

The distribution **GGS**, **RR** and **LN** can be illustrated by straight lines, if on the axis ln d, on the ordinate ln D, lnln 1/1-D (**RRB** mesh), or $\varphi(u)$ (log-normal mesh) is applied. The slope of the lines (not necessarily of the same value) gives m [33],[169],[170]; $\varphi(u)$ is the GAUSSIAN integral of errors.

It has to be noted that in the literature of cement production the slope of the lines is designated by different letters, i.e. by m at distribution **GGS** , by p at **LN**; n is called smoothing factor at **RR**. In the expressions, d_0 denotes a representative particle size. At **GGS** it means the maximum particle size, at **LN** d_0 means the particle size belonging to D=R=0.5, and at **RR** it is the so-called size module belonging to the R=1/e=0.368 which is at the same time (if the scale of d is logarithmic) the module of the distribution (see 2.11)) [169]. So for these two latter distributions, the determination of the maximum particle size which causes much trouble can be avoided.

(In our system of symbols, only m is used as the parameter of distribution, because n is reserved for denoting the dimension of the state vector generally applied in the control engineering theory of grinding).

For the comparison of the three formulae the particle size distributions of a ground cement are plotted in the same Figure (Fig. 2-2) with common abscissa and with a proper graduation of the ordinate [33].

It becomes obvious from the Figure - being in good agreement with practical experience - that the distribution **RR** gives the best approximation of the real distribution.

It can also be observed that the distribution **GGS** provides similarly good approximations for the range of fine fractions.The Hungarian practice prefers the use of the distribution **RR** since this means the best approximation of the actual particle size distribution, while the others can be regarded as special cases of the more general function (2.5). As has been shown, the distribution **RR** produces a straight line on the so-called **RRB** screen.

(However, it has to be remarked that in certain computer simulation investigations the use of the distribution **GGS** has advantages because of its very simple form; see in detail in Chapter 3).

Regarding the importance of the distribution **RR**, its other main parameters are also defined. The mean value is

$$\overline{d} = d_0 \frac{1}{m!} = \left(d_0 \frac{1}{\Gamma(m+1)} \right) \qquad (2.9)$$

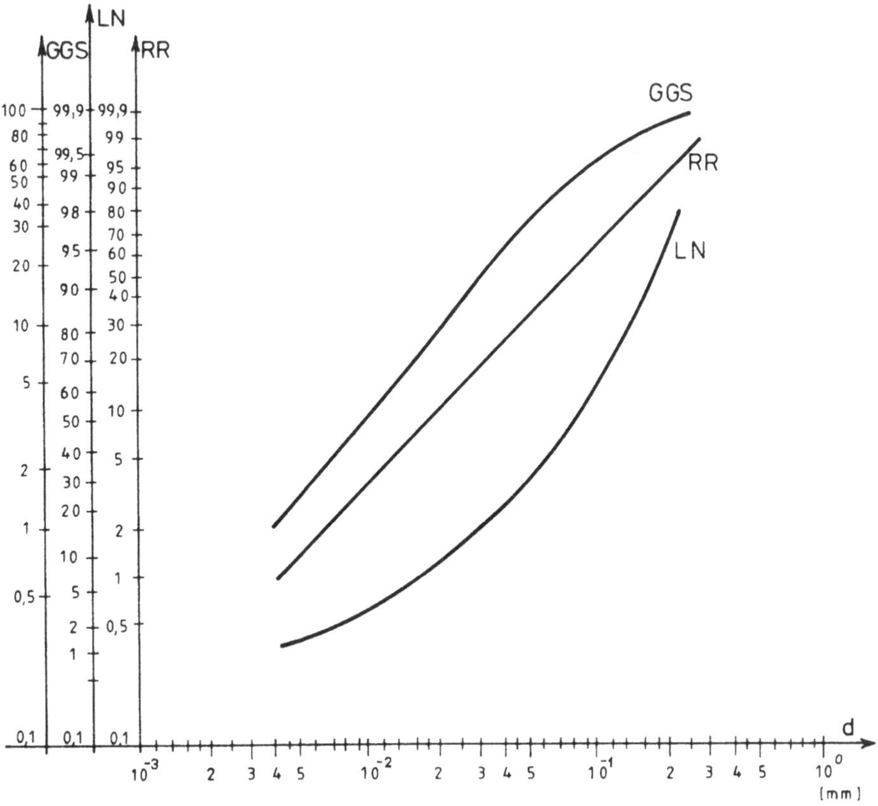

Fig. 2-2 Size distribution of the ground material plotted on **RR**, **GGS** and **LN** screens

and its deviation is

$$\sigma(d) = d_0 \sqrt{\frac{2}{m!} - \left(\frac{1}{m!}\right)^2} = d_0 \sqrt{\frac{2}{\Gamma(m+1)} - \left(\frac{1}{\Gamma(m+1)}\right)^2}, \qquad (2.10)$$

where Γ(...) is the so-called GAMMA function.

The maximum of the density or frequency function D'(d), i.e. the most probable particle size, is

$$d_M = d_0 \left(\frac{m-1}{m}\right)^{1/m} \qquad (2.11)$$

BEKE has shown in [34] that under logarithmic argument the two parameters of the **RR** definition have clear meanings, i.e. d_0 is the module of the distribution, m is obtained as the multiplication of the reciprocal of the deviation by 1.282; thus, both parameters are very important statistical measures of the stochastic process of comminution.

The expression (2.11) shows well that in practice the assumption m>1 always has to be made. Otherwise the maximum of the density function D'(d) would be obtained from a particle size of zero or negative value, which is a nonsense. However, it is a quite different matter that in the range of possible particle sizes locally a value of m<1 may give the best approximation, but under real conditions a most probable particle size different from zero must always exist, whose existence requires m>1. (Remember that m means the slope of the line if **RR** is drawn on **RRB** screen and is called the smooth factor).

<u>Definition of specific surface</u>

The ground product may be characterized by a single measure which is the surface area per unit weight, i.e. the so-called specific surface. In the knowledge of the distribution function and assuming the particles as balls, the specific surface can be calculated in a simple way as

$$S = \frac{6}{\gamma} \int_{0}^{\infty} \frac{D'(d)}{d}\, d(d) \qquad (2.12)$$

Here γ means the specific weight of the ground material. **ANSELM** in [2] proposed the following expression for the approximate calculation of the specific surface based on the **RR**-formula for the range 0.85<m<1.2:

$$S = \frac{36.8 \times 10^4}{\gamma\, m\, d_0} \qquad (2.13)$$

where d_0 is to be substituted in μm , γ in g/cm^3, and S is obtained in cm^2/g.

For finite **RR**-distribution the specific surface by (2.12) is

$$S = \frac{6}{\gamma} \frac{m}{d_0^2} \int_{d_{min}}^{d_{max}} \left(\frac{d}{d_0}\right)^{m-2} \exp\left\{-\left(\frac{d}{d_0}\right)^m\right\} d(d) \qquad (2.14)$$

where d_{max} and d_{min} mean the maximum and minimum particle sizes of the distribution, respectively. There is no explicit expression for the integral (2.14). For the case of $d_{min}=0$ and $d_{max}=\infty$, S becomes

$$S = \frac{6}{\gamma d_0} \Gamma\left(\frac{m-1}{m}\right) , \qquad (2.15)$$

which cannot be interpreted if m<1, but - as has been shown earlier - m>1 must hold true.

For the **GGS**-distribution the specific surface by (2.12)

$$S = \frac{6}{\gamma d_{max}^m} \frac{m}{m-1} \left[d_{max}^{m-1} - d_{min}^{m-1} \right] \qquad (2.16)$$

is obtained as an explicit formula having significant advantages over (2.14) from the computational aspect (see Appendix 1).

It can be concluded from the previous expressions - and this is also confirmed by measurements - that in case of the **RR**- and **GGS**-distributions the specific surface (i.e. the product Sd_0) is inversely proportional to m: it increases according to the hyperbolic or more sophisticated function by decreasing m, i.e. by decreasing the finer fractions.

Based on numerous investigations and for different feed materials **KIHLSTEDT** obtained the following empirical relationship between the specific surface and the value R(80) (retaining function value belonging to aperture size 80 μm):

$$S = \frac{k}{\gamma \sqrt{R(80)}} . \qquad (2.17)$$

In case of cement clinker, $k/\gamma = 2\times10^4$ approx. This type of equation matches well the models using retaining functions [124],[125]. Remark that **KIHLSTEDT**'s investigations were performed for wet grinding, where in general m<1. It is surprising that the square root of the retaining value is contained in **KIHLSTEDT**'s formulae, while in that of **ANSELM** it is a natural number.

Practical investigations show (see the work of **SURMANN** [210]) that even slight change in the distribution curves will cause considerable change in the value of the specific surface; accordingly, only the consideration of the complete particle size distribution can lead to accurate results and conclusions. The empirical curves based on a single point of the distribution, as for instance (2.17), may be applied only with caution.

Summary of Chapter 2

Knowledge of the particle size distributions of material flows has great importance in the investigation of grinding processes. In this chapter continuous probability functions are widely used to describe particle size distributions and their features, and validity regions are summarized on the basis of the literature. Then the computation of specific surface from distribution functions and expressions for its empirical estimate are briefly considered.

It has to be noted that the fundamentals of grinding are discussed only as far as they are needed to understand the mathematical models, as an introduction for experts not familiar with grinding technology.

For more thorough knowledge in this respect, references can be found in the text.

Symbol nomenclature in Chapter 2

$D(d)$	passing function (distribution function)
$R(d)$	retaining function
$D'(d)$	derivative of the passing function (density function)
d	particle size [μm]
d_0	a typical particle size of the distribution (e.g. maximum particle size) [μm]
φ	GAUSSIAN integral of errors
m	slope of the line representing the distribution
r	parameter of the distribution function
$\Gamma()$	the so-called GAMMA-probability function
d_M	maximum of the density function, i.e. the most probable particle size [μm]
S	specific surface [cm^2/g]
γ	specific weight [g/cm^3]
d_{max}	maximum particle size of the distribution [μm]
d_{min}	minimum particle size of the distribution [μm]
k	constant
$R(80)$	retaining function value belonging to sieve aperture (mesh) size 80 μm

3.THE STATE SPACE EQUATIONS OF BATCH GRINDING

In batch grinding, the breakage of particles can be described by the following differential equation according to the kinetic model of **LOVEDAY** [136]:

$$\frac{df(\delta)}{dt} = f(\delta) = -k(\delta)\, f(\delta), \tag{3.1}$$

where f(δ) is the fractional weight of particles of size δ, and k(δ) is the rate constant for size δ. Due to this relationship the change (decrease) in weight of particles of size δ is proportional to the weight of particles. The same assumption had been already made in the 30s and 40s in the works of the Soviet **ALYAVDIN** (1938), and the Japanese **CHUJO** (1949) [34].

Obviously the relationship (3.1) is only partly correct, because the quantity of particles of size δ increases by the crushing of particles of larger size (even in the case of batch grinding, when there is no continuous feed). The integro-differential equation of **LOVEDAY** [136] describes the above phenomenon, too:

$$\frac{df(\delta)}{dt} = f(\delta) = -k(\delta)\, f(\delta) + \int_0^{d_{max}} B(\delta,\eta)\, k(\eta)\, d\eta\,. \tag{3.2}$$

Here d_{max} means the maximum particle size of the feed, B(δ , η) is the so-called breakage (comminution) function which gives the percentage of the particles of size η that become of size δ. (However, this equation also represents the equation of conservation of the infinitely small quantity f(δ)dδ.) In practice the interpretation of size δ is difficult. Because of the problem of determining suitable equations to represent particle-size distribution, it has been convenient to use size-discretized forms of the grinding equations, where δ may denote the geometric mean of lower and upper limits of the aperture of a screen. Obviously the parameters of the discrete equation depend on the apertures of the screens and the distribution functions themselves.

Consider a set of n screens to discretize the particle size distribution where the apertures of the screens are in a geometric progression (then, with a logarithmic scale chosen, equal intervals are obtained), i.e.

$$\frac{d_i}{d_{i-1}} = a < 1 \qquad (3.3)$$

and f_i means the weight of particles falling into this interval. Accordingly, f_1 and f_2 mean the fractional weights of the coarsest and of the finest fractions, respectively. For more accurate investigations f_i is attached to the geometric mean $\sqrt{d_i d_{i-1}}$. The quotient in (3.3) is usually referred to as the sieve-ratio, too, corresponding to the measurement technique used for larger particles.

Thus, the discrete form of Eq. (3.2) given by **HORT** and **FREEH** in [102] is

$$f_i = -k_i f_i + \sum_{j=1}^{i} B(i,j)\, k_j f_j \;; \quad i=1,2,\ldots,n \qquad (3.4)$$

which means that,after crushing,the weight $B(i,j)k_jf_j$ of particles goes from the j-th fraction into the i-th fraction. The k_i values are called inner separation functions or inner-kinetic breakage constants (first-order breakage rate constants), since they mean the weight of particles retained in the interval, while the B(i,j) values are called inter-separation functions or inter-kinetic constants.

The distribution function is often taken normalized, i.e. dependent only on d_j/d_i and not on the absolute values of d_j and d_i. In this case one single vector

$$b = [b_0, b_1, \ldots, b_{n-1}]^T \qquad (3.5)$$

is sufficient to characterize B(i,j), since

$$B(i,j) = B_{i-j} \;, \qquad (3.6)$$

i.e. it is also a function of the difference i-j.
Introducing the vector

$$f = [f_1, f_2, \ldots, f_n]^T \qquad (3.7)$$

the set of first-order differential equations (3.4) can be written as a matrix differential equation

$$\dot{f} = \frac{df}{dt} = -(I-B)\,K\,f \tag{3.8}$$

where the breakage rate matrix is

$$K = \mathrm{diag}\,\langle k_1, k_2, \ldots, k_n \rangle\ , \tag{3.9}$$

the lower triangular (**TOEPLITZ**) matrix

$$B = \begin{bmatrix} b_0 & 0 & . & . & . & 0 \\ b_1 & b_0 & . & . & . & 0 \\ . & . & . & . & . & . \\ . & . & . & . & . & . \\ b_{n-1} & b_{n-2} & . & . & . & b_0 \end{bmatrix} \tag{3.10}$$

is the comminution matrix. **I** is the identity matrix.

Eq. (3.8) gives the state space equation of batch grinding (without continuous feed). This means a set of differential equations continuous in time, discretized according to the particle sizes. The dimension of the state equation is equal to the number of screens necessary to describe the distribution with a desired accuracy.

Note that **f** approximates the density function D'; therefore
$\sum_{i=1}^{n} f_i=1$, if the discretization was performed for the whole region.

The distribution function D can be easily obtained as

$$d = G\,f \tag{3.11}$$

where

$$G = \sum_{i=0}^{n-1} (S_n^T)^i\ . \tag{3.12}$$

Here

$$S_n^{\,i} = \begin{bmatrix} 0 & 0 & . & . & . & 0 & 0 & . & . & . & 0 \\ . & . & . & . & . & . & . & . & . & . & . \\ 0 & 0 & . & . & . & 0 & 0 & . & . & . & 0 \\ 1 & 0 & . & . & . & 0 & 0 & . & . & . & 0 \\ 0 & 1 & . & . & . & 0 & 0 & . & . & . & 0 \\ . & . & . & . & . & . & . & . & . & . & . \\ . & . & . & . & . & . & . & . & . & . & . \\ 0 & 0 & . & . & . & 1 & 0 & . & . & . & 0 \end{bmatrix} \qquad (3.13)$$

i-columns

n columns

$S_n^{\,i}$ is the i-th exponent of the n-dimensional shift (**TOEPLITZ**) matrix $\mathbf{S}_n$,and **d** contains the values of the passing (distribution) D(d) at particle sizes $d_0, d_1, ..., d_{n-1}$, respectively.

The character of the lower triangular matrix **B** is influenced by the fact that the crushed material will get into a finer size fraction after breakage. This corresponds to an ideal case.
By modification of **B** the state equation (3.8) can describe the case of the so-called agglomeration, too:

B =

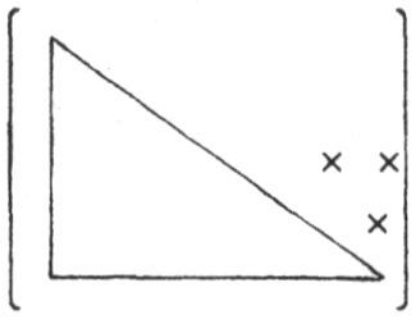

(3.14)

In this case, there are also non-zero elements in the right bottom corner of **B**; they mean that after a certain time the finer particles may stick together to form larger particles. This fact also demonstrates the wide applicability of the state equation.

The elements of the n-th row of **B** by (3.10) have a special form because the finest size fraction should have a range from d_n to zero, as a finite number of size fractions is applied. This fact includes the consideration that the elements of the n-th row are not computed according to the rule

$$b_{i,j} = b_{i-1,j-1} \qquad (3.15)$$

but are slightly modified. For an infinite number of fractions they could be computed exactly by

$$b_j' = \sum_{l=j}^{\infty} b_l \qquad (3.16)$$

Another feature has also to be mentioned. For continuous distribution the function value B(0,0) is always zero, otherwise breakage could not occur. Using a finite sieving ratio, some of the crushed material may belong to the same size interval after crushing, which can be represented by $b_0 \neq 0$. However, the determination of breakage value k'' when $b_0 \neq 0$ applied to breakage rates when $b_0=0$ (k=k') is very easy, because

$$k''(1-b_0) = k' \quad . \qquad (3.17)$$

This means, at the same time, that an ideal comminution matrix having $b_0=0$ can always be used with an appropriate breakage rate matrix **K**. (Note that the dimension of the breakage rate is 1/min !)

If the particle size distribution is known the elements of the comminution matrix can be determined under the assumption (3.6), as the size distribution function is the differential of D with respect to d, i.e.:

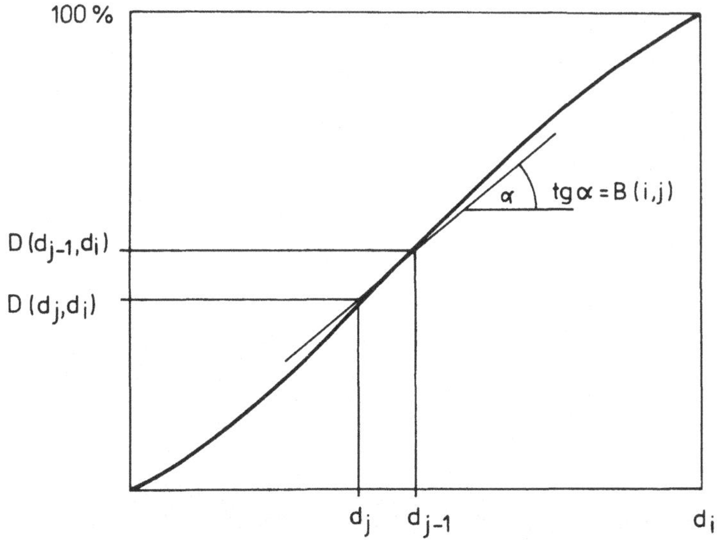

Fig. 3-1 Determination of element B(i,j) of the comminution matrix

$$B(d,d_0) = \frac{\partial D(d,d_0)}{\partial d} \quad . \tag{3.18}$$

For finite size intervals

$$B(i,j) = B_{i-j} = b_{i-j} = \frac{D(d_{j-1},d_i)-D(d_j,d_i)}{d_{j-1}-d_j} \quad , \tag{3.19}$$

as shown in Fig. 3-1.

The elements of the comminution matrix, as well as the parameters of the distribution functions may be determined by experiments and using different parameter estimation methods .

Determination of the elements of the comminution (size distribution) matrix

The elements of the size distribution matrix can be calculated only for a given distribution. Consider first the **GGS** distribution as it is quite a good approximation for fine and middle size ranges; the coarse size range has no significant role in the modelling of cement grinding. For this distribution the relationships giving the values b_i are very simple.

Using (2.2) and (3.18), the size distribution function becomes

$$B(d_i,d_0) = \frac{m}{d_0} \left(\frac{d_i}{d_0} \right)^{m-1} \quad . \tag{3.20}$$

It is shown in Appendix 2 that the elements of **B** are:

$$b_0 = 1-a^m \tag{3.21}$$

$$b_i = b_0 a^{im} \tag{3.22}$$

$$b_i{}' = a^{im} \quad . \tag{3.23}$$

These equations reduce the number of n unknown parameters in **B** to only two, i.e., the sieve ratio *a* and distribution module *m*.

Assuming **RR** distribution, the relationships (3.20)-(3.23) become more sophisticated; the calculations given in Appendix 2, however, may still be performed. Then using again (2.6) and (3.18), the size disribution matrix

$$B(d_i, d_o) = \frac{m}{d_o} \left(\frac{d_i}{d_o} \right)^{m-1} \frac{\exp \left\{ - \left(\frac{d_i}{d_o} \right)^m \right\}}{1-e^{-1}} . \qquad (3.24)$$

Compared with (3.20), it is clear, that for smaller ratio d_i/d_o (for fine fractions), both distributions provide the same approach. (Otherwise the calculation of the elements of the size distribution matrix is complicated by the fact that in (A.2.4) the partition of integrals cannot be performed due to the exponential term.)

Determination of the elements of breakage rate matrix

Prior to the examination of the transient processes of the changes in particle size distributions, the factors k_i in the breakage rate matrix **K** have to be dealt with.

As we have seen, the coefficients b_j depend essentially on the distribution and they were approximately constant for a given material to be ground. But the breakage rates k_i depend on the particle sizes, on the one hand, and are closely related with the technological and control parameters of the particular mill, on the other. Accordingly, they have significant influence on the performance of grinding.

The breakage rate has an extremum by increasing the particle size. As this maximum point belongs to particles of size 1mm, and for fine fractions dominant at cement grinding the function is nearly linear on log-log plot (see Fig. 3-2), the following expression is preferably used for characterizing the dependence on the particle sizes:

$$k_i = k_1 \left(\sqrt{\frac{d_i \, d_{i+1}}{d_0 \, d_1}} \right)^{\alpha} \qquad (3.25)$$

Selecting fractions in geometric progression (3.25) becomes simpler

$$k_i = k_1 \left[\frac{d_i}{d_o} \right]^{\alpha/2} . \qquad (3.26)$$

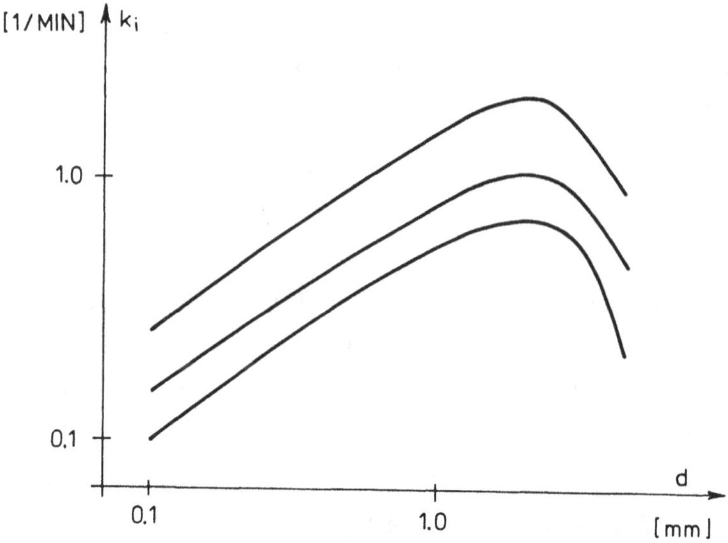

Fig. 3-2 Particle size dependence of breakage rate k_i

As the particle size distribution is discretized by a finite dimensional model, the finest particles are retained in the same interval after crushing, so k_n may be set arbitrarily, therefore a zero value is generally applied.

In ball mills the grinding is performed partly by the strike of balls moving on a trajectory and by the trimming-rubbing effect of slipping-rotating balls (cataract and cascade effect). The balls are moving on complex trajectories and there exists a critical value of the mill speed, where gravity is in balance with the radial force. In this case (at a given friction coefficient and ball filling factor) the balls adhere to the wall of the mill. This critical speed (in revolutions per minute) can be given by the diameter D of the mill

$$z_{cr} = \frac{42.3}{\sqrt{D}} \tag{3.27}$$

and the dropping altitude of balls against the mill speed is shown in Fig. 3-3, i.e. it has a maximum at

$$z_{opt} = \frac{32}{\sqrt{D}} \tag{3.28}$$

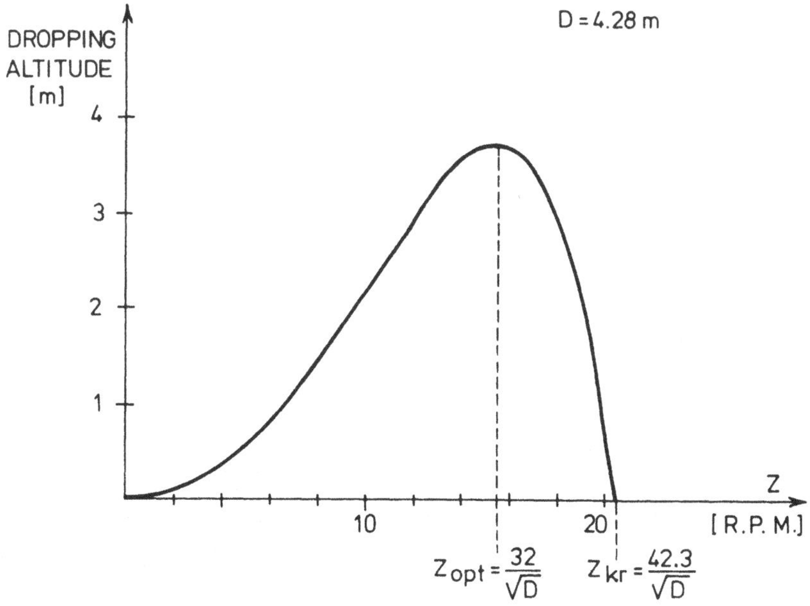

Fig. 3-3 The dropping altitude of the balls with respect to the r.p.m. of the mill

(The ball mills are operated usually in this flat maximum region [33].)
The fact is closely related to the phenomenon that the breakage rates k_i also have their maxima as a function of mill speed (see Fig. 3-4). Here

$$z = \frac{Z}{Z_{cr}} \tag{3.29}$$

is used, as the further relationships are valid only up to the critical speed.
The optimum speed values in the two Figures do not necessarily coincide, since the effectiveness of grinding depends not only on the dropping altitude but on several other factors. It has to be noted that the curve $k_i(z)$ is not yet zero at z=1, since theoretically, at this crucial situation, there still exists grinding not resulting from dropping of the balls.
The breakage rate is dependent on the ball diameters d_B, too, and for different fractions it can be given by curves having maxima as shown in Fig. 3-5.
Similar curves having maxima can be given to demonstrate the dependence of the breakage rate on

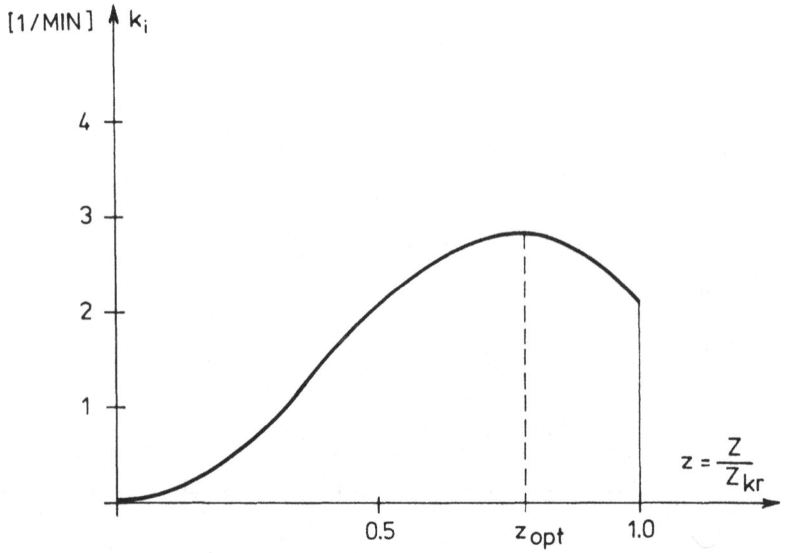

Fig. 3-4 The breakage rate k_i with respect to the r.p.m. of the mill

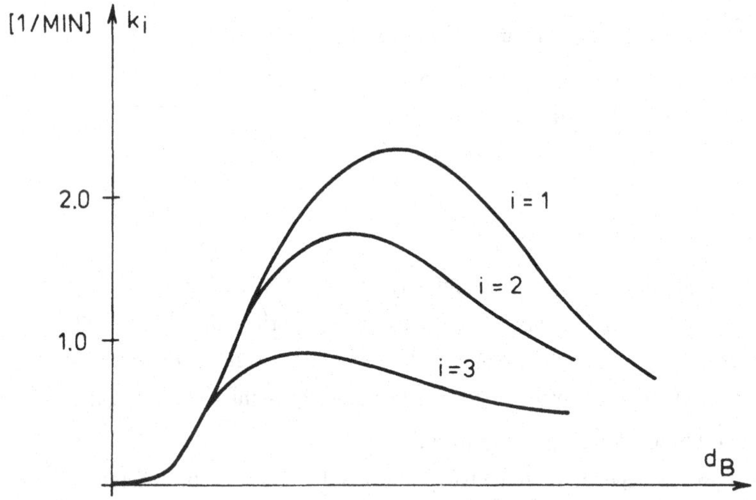

Fig. 3-5 The breakage rate with respect to the ball diameter d_B

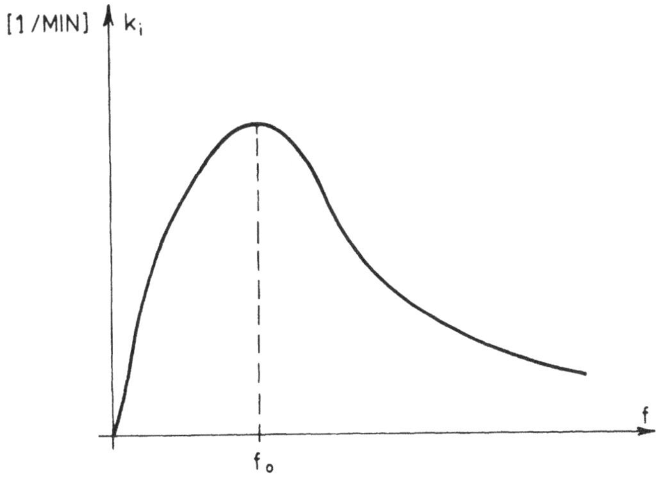

Fig. 3-6 Dependence of the breakage rate on the material f being in the mill

the load of material f in the mill (or filling factor), see Fig. 3-6. When the mill charge is large, k_i is inversely proportional to f with approximately hyperbolic character.

At decreasing charges, when the mill charge fills approximately the interstitial volume of the static ball charge, the effectiveness of grinding declines, because collisions are mainly between the balls, on the one hand, and the specific value of the energy exceeds that necessary to cause agglomeration, on the other. (This fact is in good agreement with experience, according to which the module *m* of **RR** distribution is dependent on the mill charge and has maximum value for batch (and open circuit) grinding.)
Considering the above behaviours, the breakage rates k_i can be given by the following complex formula [154] (in good agreement with experience):

$$k_i = k_{io} \frac{d f z}{(f_o^2 + f^2) d_B^2} \exp\left\{ \frac{-\gamma_i}{d_B^3 z^2 (1 + \beta_1 z^4 + \beta_2 z^8)} \right\} . \qquad (3.30)$$

It is reasonable to set the value of k_{io} to the value, dependent on the particle sizes, which ensures the same conditions as (3.26) under constant parameters and variables. A possible choice may be

$$k_i = k_i^* \frac{d_i^\alpha f z}{(f_o^2 + f^2) d_B^2} \exp\left\{ \frac{-\gamma_i}{d_B^3 z^2 (1 + \beta_1 z^4 + \beta_2 z^8)} \right\} \qquad (3.31)$$

whose derivation is given in Appendix 3.

In practice the ball diameters are different (in fact, uniform ball size is not at all desirable), so the above expression can be regarded as a general one valid for an average ball size within fractions or a group of fractions (we refer to the ball mills of two compartments filled, of course, with balls of different sizes).

Solution of the state equations

Returning to the state equation (3.8), it is well suited for modelling the dynamics of size distribution and to the use of different simulation techniques. The solution of the homogeneous vector difference equation is obtained as

$$f(t) = e^{(B-I)Kt}\, f(0) \tag{3.32}$$

for an initial particle size distribution, as is well known from the theory of linear systems [48]. As far as k_i and k_j are different (which is the general case), we find the analytical solution

$$f(t) = A\, J(t)\, A^{-1} f(0), \tag{3.33}$$

where

$$A_{ij} = \begin{cases} 0 & , \text{ if } i > j \\ 1 & , \text{ if } i = j \\ \sum_{l=j}^{i-1} \dfrac{b_{il} k_l}{k_i - k_j} A_{il} & , \text{ if } i < j \end{cases} \tag{3.34}$$

and

$$J_{ij} = \begin{cases} 0 & , \text{ if } i \neq j \\ e^{-k_i t} & , \text{ if } i = j \end{cases} \tag{3.35}$$

It is useful, from the computational aspect, to discretize the state equation in time, too, with equidistant sampling time h, and to apply the vector difference equation [48]

$$f(t+1) = \Phi(h)\, f(t) \quad . \tag{3.36}$$

Here

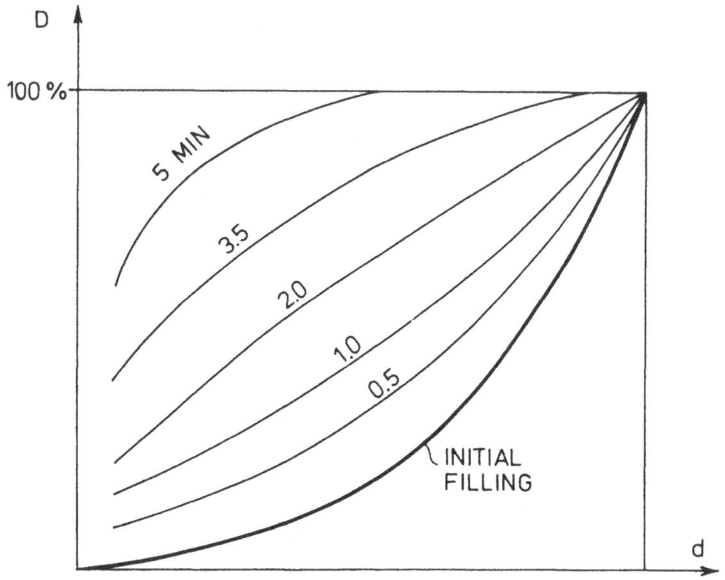

Fig. 3-7 Dynamic behaviour of the particle size structure (for the **GGS** distribution)

$$\Phi(h) = e^{(B-I)Kh} \tag{3.37}$$

The solution of the state equation gives the dynamic behaviour of particle size structure (for the **GGS** distribution) as it is seen in Fig. 3-7. By means of this the desired grinding time (very important in the case of agglomeration!) can be easily determined.

The steady-state particle size distribution matrix model introduced by **BROADBENT-CALCOTT** in [42] is well known in the literature; it gives the expression of the size distribution $f(t_g)$ at the end of the grinding time t_g, assuming feed size distribution $f(0)$

$$f(t_g) = (I-\tilde{K}+\tilde{B}\tilde{K})\ f(0) \tag{3.38}$$

and its modified version is also known as

$$f(t_g) = (I-\tilde{K}+\tilde{B}\tilde{K})^{\nu} f(0) \tag{3.39}$$

According to **LYNCH** , ν is proportional to the residence time in the mill of the material, i.e. to grinding time t_g.

However, these equations are only rough approaches to the above shown dynamic solutions and have slight applicability.

On the basis of (3.32) we may write

$$f(t_g) = e^{(B-I)Kt_g} f(0) = (e^{(B-I)K'})^{\nu} f(0) , \qquad (3.40)$$

where $k_i = a_{11} k_i'$ and $\nu = a_{11} t_g$; furthermore, $(I - \tilde{K} + \tilde{B}\tilde{K})$ as well as $\exp\{(\mathbf{B}-\mathbf{I})\mathbf{K}'\}$ are lower triangular matrices, but the letters have quite different meanings. It is seen that ν is in reality proportional to the grinding time.

In the special case when $\tilde{B} = B$, $\tilde{K} = K'$, the Taylor series of the exponential function may be written as

$$e^{(B-I)K'} = I + (B-I)K' + \frac{1}{2!} [(B-I)K']^2 + \ldots \qquad (3.41)$$

If $(b_0-1)k_i << 1$ for all i, then

$$e^{(B-I)K'} \approx I + (B-I)K' \qquad (3.42)$$

and

$$f(t_g) \approx (I - K' + BK')f(0). \qquad (3.43)$$

This derivation is detailed in order to demonstrate that, although grinding theory is mainly based on empirical relationships, certain phenomena can be modelled not only by experiments; this applies mainly when the particular process can be simply interpreted by the theory of linear systems. This simple example also demonstrates the necessity of an advanced, complex, system and control engineering revision of the theory of grinding.

It should be mentioned in connection with the previous comments that, based on experience, the change in time of the particle size distribution is considered according to the expression

$$D(d,t) = [1-e^{-qt^p}][D(d,\infty)-D(d,0)] \qquad (3.44)$$

which also gives a straight line on log-log plot [81]. The value of p is estimated in the literature to be in the range 0.7 - 1.5.

It is well known from the theory of linear dynamic systems that due to (3.32) the system response for one fraction has the same character as that of (3.44) regarding the magnitude of the eigenvalues of matrix **(B-I)K** and their relative positions. Thus the time function of the size distribution can be investigated by the classical test methods of linear dynamic systems (eigenvalues, Laplace transformation, etc.).

Very sophisticated control problems are capable of solution by the model of batch grinding introduced in this chapter. For example, the optimum grinding time can be determined from the time function of the size distribution. With known grinding time, the chosen control variables (e.g. mill speed, ball size, etc.) can be optimized in order to maximize the fine product $f_n(t_g)$. An optimum product density function $\mathbf{f}^*(t_g)$ can be prescribed in advance which should be reached as closely as possible by the end of the grinding time. For this purpose the quadratic cost function

$$J = [f(t_g)-f^*(t_g)]^T W [f(t_g)-f^*(t_g)] \qquad (3.45)$$

has to be minimized [154]. Here the element w_i in the main diagonal of weight matrix **W** is proportional to the relative importance of the i-th fraction. This principle can be extended to the simultaneous minimization of the grinding time.

Connection between the classical theory of grinding and the state equations

Early investigations - regarded today as classical ones - aimed at a better understanding of the comminution process were concerned with the relationship between the energy consumed by a grinding mill and the extent of size reduction that the consumption of this energy brought about, as well as with the diameter of the product particles.

According to **RITTINGER**'s law from 1867, the energy consumed is proportional to the difference between the feed and product surface area of particles [181].

KICK's law from 1883 states that the energy consumption per unit weight is constant [33].

The much newer relationship proposed by **BOND** (from 1952) says that the energy consumed is proportional to the difference between the reciprocal values of the square roots of the feed and product particle sizes [40].

It is obvious that the three laws cannot hold true simultaneously. Experiments showed that **RITTINGER**'s law gives quite a good approximation for small particles, while that of **KICK** is better for large particles. **BOND**'s law is, however, valid for particles of medium sizes.

The three laws may be expressed simply by the equation

$$\frac{dE}{d(d)} = - \mu \frac{1}{d^{\nu}} , \qquad (3.46)$$

where E is the energy consumed, μ and ν are constants. By substituting $\nu=2$ and integrating (3.46), we get **RITTINGER**'s law; for $\nu=1.5$, **BOND**'s law; and for $\nu=1.0$ **KICK**'s law is obtained [154].

The relationship of the specific surface is defined in (2.12) and can be rewritten as

$$S = \frac{6}{\gamma} \int_0^{\infty} \frac{D'(d)}{d} d(d) = \frac{6}{\gamma d_0} \int_0^{1} \frac{D'(d/d_0)}{d/d_0} d(d/d_0) = \frac{6 s_a}{\gamma d_0} , \qquad (3.47)$$

where s_a is a *shape* factor resulting from integration, and a normalized size distribution function is used (i.e. dependent only on d/d_0).

Thus the specific surface is inversely proportional to d_0, which is usually chosen as the upper limit for the sizes of particles (i.e. for d/d_0, the upper limit of integration becomes 1).

By the fundamental equation of comminution (3.1), at first sight the following quantity of crushed material derives from unit weight during unit interval, having the distribution as a set of n screens, i.e.

$$\sum_{i=1}^{n} df_i = \sum_{i=1}^{n} k(d_i) f_i \quad . \qquad (3.48)$$

If the feed surface is negligible compared with the product surface, then the new surface area produced is

$$\Delta S = \frac{6 s_a}{\gamma} \sum_{i=1}^{n} k(d_i)\, f_i \frac{1}{d_i} \quad . \tag{3.49}$$

Here (3.47) is used, and d_i means the apertures of screens compared with the maximum value d_0. Taking $k(d_i)$ proportional to d_i, i.e.

$$k(d_i) = a_1 d_i \quad , \tag{3.50}$$

and assuming constant energy, we get

$$\Delta S_1 = \frac{6 s_a a_1}{\gamma} = \text{constant} \tag{3.51}$$

which, under the assumptions made, gives **RITTINGER**'s law.

Taking $k(d_i)$ independently from d_i, i.e.

$$k(d_i) = a_2 = \text{constant} \quad , \tag{3.52}$$

and according to (3.48), the quantity of crushed material obtained from unit weight during unit grinding time is constant - which is the statement of **KICK**'s law. Taking $k(d_i)$ to be proportional to the square root of d_i, i.e.

$$k(d_i) = a_3 \sqrt{d_i} \quad , \tag{3.53}$$

we get

$$\Delta S_3 = \frac{6 s_a a_3}{\gamma \sqrt{d}} \quad , \tag{3.54}$$

which corresponds to **BOND**'s law.

(Actually the above mentioned features can be seen from the different form of curve k(d) to be valid for small, medium and large particles.)
Of course, all that has been said serves the purpose of demonstrating that the state equation, introduced here, by an appropriate choice of k(d), may include the results of the classical theory of grinding and is not in contradiction with it.

Calculation of the specific surface

Applying discrete density vectors, the integral in (2.12) can be evaluated easily by taking finite summa instead of the integral:

$$S = \frac{6}{\gamma} \sum_{i=1}^{n} \frac{f_i}{d_i} \quad . \tag{3.55}$$

Having apertures in geometrical progression ($d_i = a^i d_0$):

$$S = \frac{6}{\gamma d_0} \sum_{i=1}^{n} f_i \, a^{-i} \quad . \tag{3.56}$$

The expression of the specific surface for the **GGS** distribution is deduced in Appendix 1, i.e.

$$S = \frac{6}{\gamma d_0} \frac{m}{m-1} \left[1 - a^{n(m-1)} \right] \quad . \tag{3.57}$$

(Unfortunately, for the **RR**-distribution the integral by (2.12) cannot be expressed in an analytical form.)

Summary of Chapter 3

For mathematical modelling of batch grinding, the kinetic model of **LOVEDAY** is applied. By discretizing the particle size distribution and density functions of continuous grinding with a set of screens into a finite histogram, the discrete version of the integro-differential equation of the grinding process is obtained. Rewriting it into a matrix form, the state equation characterizing the dynamic particle size distribution relations of batch grinding is determined. Then the phenomenon of agglomeration is treated, and the determination of the elements of comminution matrix and breakage rate matrix and their relation to the parameters of the technology, are discussed.

The solution of the state equation is presented, and it is proved that the time function of the size distribution can be investigated by the classical test methods of linear dynamic systems.

Finally the connection between the classical theory of grinding and the state equations is considered; then the computation of the specific surface from discrete data is given.

Symbol nomenclature of Chapter 3

δ	particle size (diameter) [μm]
$f(\delta)$	fractional weight of particles of size δ
$\dot{f}(\delta) = \dfrac{df(\delta)}{dt}$	
$k(\delta)$	breakage rate [1/min]
$B(\delta,\eta)$	comminution (size distribution) function
f_i	fractional weight of particles in the size interval i
k_i	inner-separation function
$B(i,j)$	inter-separation function
$\mathbf{f}$	vector of fractional weights
$\mathbf{K}$	breakage rate matrix
$\mathbf{B}$	comminution matrix
$\mathbf{I}$	identity matrix
$\mathbf{G}$	auxiliary matrix
$\mathbf{S}_n$	n-dimensional shift (**TOEPLITZ**) matrix
a	sieve ratio
Z	mill speed (revolutions per minute: r.p.m.)
Z_{cr}	critical mill speed
Z_{opt}	optimum mill speed
z	relative mill speed
D	inside diameter of the mill [m]
D_B	ball diameter [cm]
f	feed
f_o, β_1, β_2	parameters unaffected by the particle sizes
k_i^*, γ_i	parameters dependent on the particle sizes
$\mathbf{A}$	transformation matrix

$\mathbf{J}(t)$	JORDAN matrix
$\Phi(h)$	auxiliary matrix
ν	constant
$\mathbf{K}'$	modified breakage rate matrix
t_g	grinding time
$\mathbf{W}$	diagonal weighting matrix
E	energy consumed
μ	constant
s_a	shape factor
a_1,a_2,a_3	contants

4. MODELLING OF OPEN CIRCUIT GRINDING

First consider an idealized *one-dimensional* mill model of open circuit or continuous grinding (see Fig. 4-1). Here x(t), m(t) are the material flows of feed and mill product, respectively. Let **x** and **m** denote the discrete distributions approximating to the density functions of size distributions of the two material flows according to (3.7), whose i-th elements give the probability of falling into the proper interval. Let the so-called summing vector be

$$\mathbf{1} = [1, 1, \ldots, 1]^T \tag{4.1}$$

whose scalar product by the above vectors always gives unity (assuming that the entire size interval is included), which is obvious from the definition of the density function. Thus $\mathbf{1}^T\mathbf{x}=1$, etc.

Let f(t) be the total volume in the mill and **f**(t) be the density function of the size distribution of this material. Distributed parameter models, based on convective material flows and equations of mixing, were first suggested to describe continuous grinding. These models, however, could include only very simple schemes as in Fig. 4-2, i.e. when the radial and tangential variations are regarded as negligible compared with the axial variations, the model was *one-dimensional.*
Let the mill length be L.

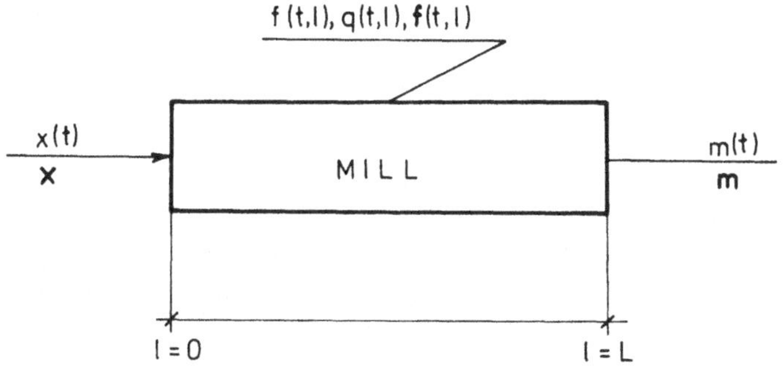

Fig. 4-1 Ideal, *one-dimensional* mill

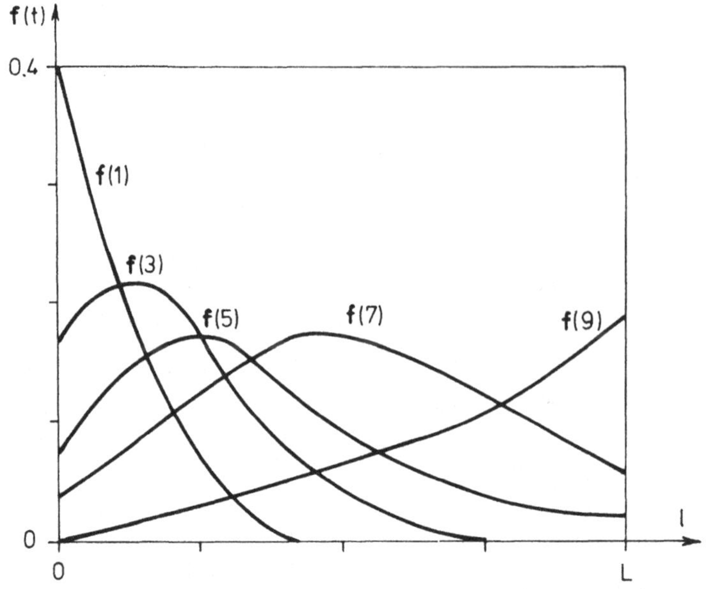

Fig. 4-2 Particle size distributions along the mill in the function of time

A mass balance for an infinitesimal element Δl along the mill gives the partial differential equation [158]:

$$f(t,l)\ \frac{\partial f(t,l)}{\partial t} + q(t,l)\ \frac{\partial f(t,l)}{\partial l} =$$

$$= E\ \frac{\partial^2[f(t,l)f(t,l)]}{\partial l^2} + f(t,l)\ (I-B)K\ f(t,l) \qquad (4.2)$$

E is the coefficient matrix of second order dispersion (in simple cases it is a diagonal matrix), q is the material flow along the mill. Here f(t,l) is regarded as the mass per unit length.

The other quantities are already known from the results obtained for batch grinding in Chapter 3. It is seen that all variables, even **E**, **B** and **K**, depend on time and space. As a first approximation the dependence of these matrices on time and space may be eliminated. Then the partial vector differential equation has to be solved under boundary conditions:

$$x(t)x = q(t,0)f(t,0) - E\ \frac{\partial[f(t,0)f(t,0)]}{\partial l} \qquad (4.3)$$

and

$$V q(t,L) f(t,L) = m(t) m(t) \quad . \tag{4.4}$$

V is the so-called outlet classification matrix which will be discussed later in this chapter.

At the outlet, the condition

$$\frac{\partial f(t,L) f(t,L)}{\partial l} = 0 \tag{4.5}$$

must also hold true, as the grinding is finished there. Let us suppose that the material distribution is uniform in the mill, i.e. f(t,l)=f(t), then (4.2) can be modified according to

$$\frac{\partial f(t,l)}{\partial t} + \frac{1}{\tau(t,l)} \frac{f(t,l)}{\partial l} = E \frac{\partial^2 f(t,l)}{\partial l^2} + (I-B)Kf(t,l) \tag{4.6}$$

where

$$\tau(t,l) = \frac{f(t)}{q(t,l)} \tag{4.7}$$

is the average residence time depending on space. In the solution of (4.6) the dispersion term is the source of difficulty. Usually this term is neglected as the estimation of **E** is rather complicated. It is seen that this term has influence only on the high-frequency behaviour of the model, and does not affect significantly the operational region of the plant.

By neglecting this term,and by introducing the time and space independent fraction speed v_i (the speed of particles in the fraction i relative to an average speed or to the speed of transport material), the distributed parameter model of continuous grinding introduced by **FUERSTENAU** and **MIKA** [87],[88] can be obtained:

$$\frac{\partial f(t,l)}{\partial t} + V \frac{\partial f(t,l)}{\partial l} = (I-B)Kf(t,l) \quad , \tag{4.8}$$

where **V** is a diagonal matrix

$$V = \mathrm{diag} \langle v_1, v_2, \ldots, v_n \rangle \tag{4.9}$$

and

$$v_i = \frac{1}{\tau_i} = \frac{q_i(t)}{f(t)} = \frac{q_i}{f} \quad . \tag{4.10}$$

The above model equation, assuming a step excitation function in f(0,0), can also be solved by Laplace transformation, but it is reasonable to make discretizations in time and in axial distances along the mill. The curves in Fig. 4-2 are obtained by computer simulation using this kind of model, where the particular fractions are presented along the mill, and it can be observed that the finer fractions are in larger quantity at the end of the mill.

Thus, the distributed parameter models for continuous grinding present considerable difficulties ,in spite of the neglected terms, and can be used in the main only for simulation purposes. For use in control systems, a reduced model which gives a satisfactory dynamic and stationary description of the process is desirable.Such a model is given in the following, using the results of Chapter 3.

Considering again Fig. 4-1, the total mass balance of the mill is

$$\frac{df(t)}{dt} = x(t) - m(t) \; . \qquad (4.11)$$

The mass balance for each size fraction is

$$\frac{df(t)\mathbf{f}(t)}{dt} = (\mathbf{B}-\mathbf{I})\mathbf{K}f(t)\mathbf{f}(t) + x(t)\mathbf{x}(t) - m(t)\mathbf{m}(t) \; . \qquad (4.12)$$

This equation is in full agreement with (4.2), only the dispersion is neglected and differentials by l are replaced by differences along the axis of the mill. Thus f(t) means the total volume in the mill, **f**(t) representing its size distribution function.

Let us introduce the vector variables denoted by ', whose components give the volumes in the fractions and can be calculated as products of *specific* size distribution vectors and material flows, e.g. $\mathbf{m}'(t)=m(t)\mathbf{m}(t)$.

Using the notations introduced above, (4.12) becomes

$$\frac{d\mathbf{f}'(t)}{dt} = (\mathbf{B}-\mathbf{I})\mathbf{K}\mathbf{f}'(t) + \mathbf{x}'(t) - \mathbf{m}'(t) . \qquad (4.13)$$

Thus, f(t) and **f'**(t) or f(t) and **f**(t) represent the state variables in the dynamic size distribution model of continuous grinding.

The integral form of the dynamic mass balance equation (4.11)

$$f(t) = f(0) + \int_0^t [x(\tau) - m(\tau)]d\tau \qquad (4.14)$$

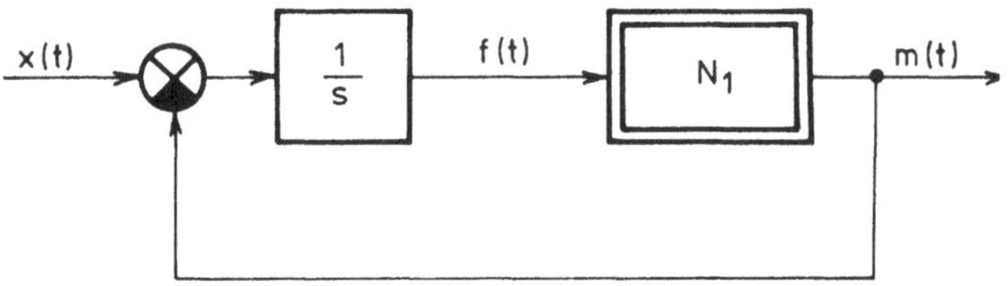

Fig. 4-3 Dynamic control engineering model of the mill for describing quantitative relations

leads to a very simple closed loop control system, shown in Fig. 4-3. The nonlinearity N_1 represents the relationship between the mill outlet m(t) and the total volume in the mill. In the usual operational region a linear approach

$$m(t) = N_1[f(t)] \approx N_1\ f(t) \tag{4.15}$$

is satisfactory, but for more accurate investigations a nonlinear curve has to be taken into account (see Section 5.1).

Multiply (4.13) by $\mathbf{1}^T$ from the left side

$$\frac{d1^T f'(t)}{dt} = \frac{df(t)}{dt} = [1^T(B-I)Kf(t)]f(t) +$$

$$+ 1^T x(t)x(t) - 1^T m(t)m(t) = x(t)-m(t) \ , \tag{4.16}$$

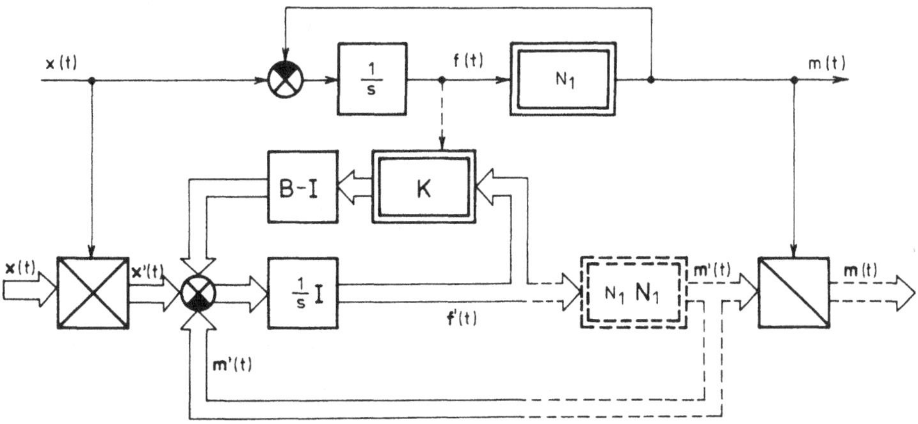

Fig. 4-4 Control engineering model for describing the quantitative and particle size distribution relations of the mill

as if **x**(t), **m**(t), **f**(t) represent the whole region of particle sizes, then their product by 1^T produces unity. For similar reasons the first term of the right-hand side becomes zero. Thus (4.11) and (4.12) are not contradictory at all. The two basic relationships are illustrated also in Fig. 4-4. The upper part shows the block-schemes of (4.11) and (4.15), and the lower part that of (4.12). The new task is to establish the relationships for the part indicated by dotted line.

Eq. (4.13) can be easily integrated as, on the left-hand side, there are the changes of mass flows belonging to different fractions; the right-hand side, however, shows that these changes partly result from grinding (see the first term, which completely corresponds to the description obtained for batch grinding), partly from the difference between the feed and outlet.

Suppose - as a generalization of (4.15) - that the same relationship exists between the size distributions of the outlet from, and volume in, the mill, i.e.

$$m(t) = N_1 \; f(t) \; . \tag{4.17}$$

The meaning of $\mathbf{N}_1$ will be discussed later. Applying the above expressions, the dependence of **m**'(t) on **f**'(t) can be determined as follows:

$$m'(t) = m(t)m(t) = m(t) \; N_1 \; f(t) = m(t) \; N_1 \; \frac{1}{f(t)} \; f'(t) =$$

$$= \frac{m(t)}{f(t)} \; N_1 f'(t) = N_1 N_1 f'(t) \; . \tag{4.18}$$

This refers to the part of Fig. 4-4 drawn in dotted line. Since f(t) and **f**(t) are independent state variables, it seems to be reasonable to find an equation, instead of (4.13), which has only **f**(t) as a state variable.

From (4.13) we obtain

$$\frac{df(t)f(t)}{dt} = \frac{df(t)}{dt} \; f(t) \; + \; f(t) \; \frac{df(t)}{dt} = \frac{df'(t)}{dt} \; , \tag{4.19}$$

whence by using (4.11)

$$\frac{df(t)}{dt} = \frac{1}{f(t)} \; [\; \frac{df'(t)}{dt} - \frac{df(t)}{dt} \; f(t)] = (B-I)Kf(t) \; +$$

$$\frac{1}{f(t)} \; \{[x(t)x(t)-m(t)m(t)]-[x(t)-m(t)]f(t)\} . \tag{4.20}$$

Finally, taking (4.17) into consideration,

$$\frac{d\mathbf{f}(t)}{dt} = \{(\mathbf{B}-\mathbf{I})\mathbf{K}-[\frac{x(t)}{f(t)} - N_1]\mathbf{I}-N_1\mathbf{N}_1\}\mathbf{f}(t) + \frac{x(t)}{f(t)}\mathbf{x}(t) \qquad (4.21)$$

the nonlinear state equation having varying parameters is obtained, yielding - together with (4.11) - the dynamic size distribution model of continuous grinding.

Let us observe that the inputs of the model are x(t) and **x**(t) - in reality -, while the outputs may be m(t) and **m**(t). The state variables are f(t) and **f**(t). In steady-state, df(t)/dt=0, so x(∞)=m(∞) and d**f**(t)/dt=**0**, thus

$$[(\mathbf{I}-\mathbf{B})\mathbf{K}+N_1\mathbf{N}_1]\mathbf{f} = N_1\mathbf{x} \quad , \qquad (4.22)$$

where $x(\infty)/f(\infty)=m(\infty)/f(\infty)=N_1$ is taken into account. The steady-state (average) size distribution in the mill is

$$\mathbf{f} = [\frac{1}{N_1}(\mathbf{I}-\mathbf{B})\mathbf{K}+\mathbf{N}_1]^{-1}\mathbf{x} \quad , \qquad (4.23)$$

while the size distribution of the mill outlet is

$$\mathbf{m}=\mathbf{N}_1\mathbf{f}=\mathbf{N}_1[\frac{1}{N_1}(\mathbf{I}-\mathbf{B})\mathbf{K}+\mathbf{N}_1]^{-1}\mathbf{x}=[\mathbf{I}+(\mathbf{I}-\mathbf{B})\mathbf{K}(N_1\mathbf{N}_1)^{-1}]^{-1}\mathbf{x}. \qquad (4.24)$$

(It has to be remarked, as a curiosity, that these equations have the same structure as the **BROADBENT-CALCOTT** model in [42], but these are based on theoretical investigations and not on experimental and empirical results.)

Consider now the separation matrix $\mathbf{N}_1$ at the outlet. The residence time τ_i for size i depends on the material being in the i-th fraction and the outlet:

$$\tau_i = \frac{f_i'}{m_i'} = \frac{f(t)f_i}{m(t)m_i} = \frac{f(t)}{m(t)}\frac{f_i}{m_i} = \tau\frac{f_i}{m_i} = \frac{1}{U_i} \quad , \qquad (4.25)$$

where

$$\tau = \frac{f(t)}{m(t)} = \frac{1}{U} \qquad (4.26)$$

is the mean residence time. V_i and V denote the fractional speed and mean speed, respectively.

It follows from (4.25) that the weight ratio in the fraction i depends - through the residence time - on the i-th fractional weight rate in the mill, i.e.:

$$m_i = \frac{\tau}{\tau_i} f_i = \frac{V_i}{V} f_i \, . \qquad (4.27)$$

Comparing (4.17) and (4.27), we get

$$N_1 = \begin{bmatrix} \left(\frac{\tau_1}{\tau}\right)^{-1} & \cdots & 0 \\ & & \\ 0 & \cdots & \left(\frac{\tau_n}{\tau}\right)^{-1} \end{bmatrix} = \begin{bmatrix} \frac{V_1}{V} & \cdots & 0 \\ & & \\ 0 & \cdots & \frac{V_n}{V} \end{bmatrix} = V$$

(4.28)

i.e. the outlet separation is a diagonal matrix; its elements are inversely proportional to the relative residence time of the fractions or proportional to the relative speeds. (Note that theoretically any other quantities can be chosen as reference (see later) e.g. the speed of transporting air or water flow.)

It is obvious that the choice $\mathbf{N_1}=\mathbf{I}$ would have been incorrect, as then it would have been necessary to use breakage rates different from those obtained in batch grinding which would be inconsistent with reality. Now it can be better understood why the velocity matrix has been used in (4.8). Thus the relative fractional speeds

$$v_i = \frac{V_i}{V} \qquad (4.29)$$

have a very important role in the dynamics of continuous grinding. According to experience, the speed of the finest fractions is the highest , which means that the residence time of the coarsest particles is the longest. Consequently, the relative axial velocity depends on the particle size approximately according to Fig. 4-5. For simulation purposes the formula

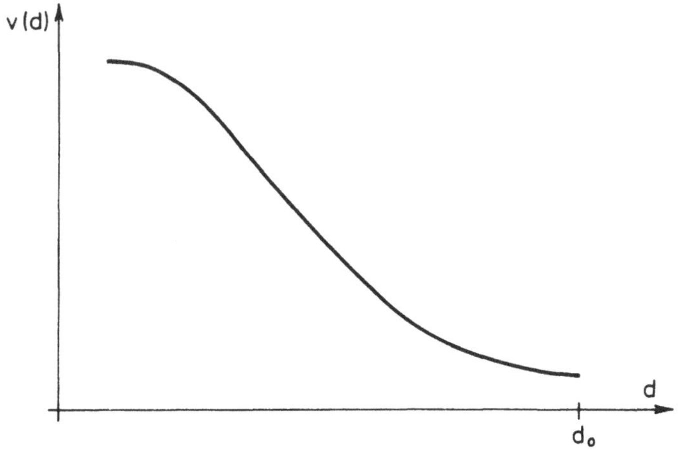

Fig. 4-5 Dependence of the relative axial speed on the particle size

$$v_i = \left(\frac{d_n^2}{d_i d_{i-1}}\right)^{\beta} \tag{4.30}$$

is generally used, whose simpler form can be obtained for screens having apertures in geometric progression:

$$v_i = \left(\frac{a d_n^2}{d_i d_i}\right)^{\beta} = a^{\beta}\left(\frac{d_n}{d_i}\right)^{2\beta} \tag{4.31}$$

Instead of this, however, it is reasonable to use the formula

$$v_i = \left(\frac{d_n}{d_i}\right)^{2\beta} \tag{4.32}$$

as the finest fraction to leave the mill with the transporting material (air or water); thus the normalization $v_n=1$ can be performed. This is shown in Fig. 4-6.

The relative fraction velocities v_i do not depend in practice on the grinding condition parameters, so they can be regarded as constants.

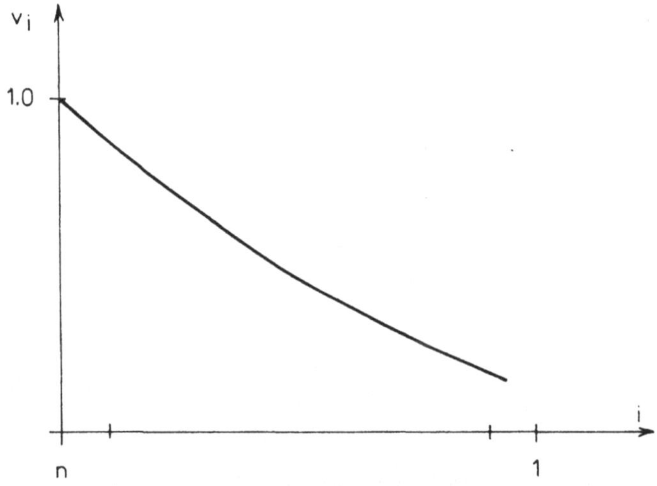

Fig. 4-6 Relative speed of particular size fractions

In the model the following assumptions were made:

- the form of the function v(d) is constant,
- the relative ratio of the values v_i is also constant.

However, due to the indestructibility of matter, it is necessary to change the absolute value of v_i, following from (4.17), in order to hold equality

$$m(t) = 1^T m'(t) = m(t) 1^T m(t) = m(t) 1^T N_1 f(t) \; . \qquad (4.33)$$

Accordingly, for the elements of $\mathbf{N}_1$ there are further restrictions depending also on $\mathbf{f}(t)$,

$$1^T N_1(t) f(t) = \sum_{i=1}^{n} v_i f_i = 1 \qquad (4.34)$$

which transforms the virtually linear operator $\mathbf{N}_1$ into a nonlinear one. This can also be achieved by applying the actual quantities

$$v_i^* = \frac{1}{\sum_{i=1}^{n} v_i f_i} v_i = \frac{v_i}{1^T N_1 f(t)} \qquad (4.35)$$

instead of v_i determined in advance.

Thus

$$N_1 = \frac{1}{1^T U f} U = \mathrm{diag} \langle v_1^*, v_2^*, \ldots, v_n^* \rangle \ . \qquad (4.36)$$

(Note that since $\mathbf{N}_1$ is a function of $\mathbf{f}(t)$, the steady-state relations developed here can be regarded as approximative only. To obtain an accurate solution, a set of nonlinear equations must be solved.)

In the foregoing, when modelling the residence time of the fractions, the axial fraction velocity was related to a fictitious "mean" speed m(t)/f(t) by (4.26). We have seen that in this case certain manipulations were required (see normalization by (4.35)) to ensure the mass balance. This resulted in a special nonlinearity for $\mathbf{N}_1$, which avoids difficulties from the computational aspect, but the consequence is that N_1=constant. This fact, however, as experience has shown, does not hold true for cement mills; see the related statement of section 5.1. (It is worth considering, however, whether this is the correct mode of description in wet grinding.) Applying a mean velocity as a reference, the principles of normalization related to Fig. 4-6 cannot be fulfilled either, because there exists a fraction of still higher velocity in the flow.

Let us consider another approach assuming, that the transporting material is air. For the residence time of the i-th fraction in the mill, (4.25) is obviously a correct expression, so

$$f_i' = \tau_i m_i' = \frac{L}{V_i} m_i' = \frac{L}{V_0} \frac{V_0}{V_i} m_i' = \frac{L/V_0}{V_i/V_0} m_i' = \frac{\tau_0}{v_i} m_i' , \qquad (4.37)$$

where L is the length of the mill, V_0 the velocity of the transporting air flow and τ_0 the residence time of the air. Assuming that the finest fraction leaves the mill with a velocity the same as that of the air flow, the relative velocities

$$v_i = \frac{V_i}{V_0} \qquad (4.38)$$

have a physical meaning.

From (4.37) we obtain

$$m'(t) = \frac{1}{\tau_0} Uf'(t) = N_1 f'(t) \qquad (4.39)$$

where now

$$N_1 = \frac{1}{\tau_0} U \qquad (4.40)$$

differs from what has been used until now and **V** is unchanged. τ_0 can be easily calculated

$$\tau_0 = \frac{L}{U_0} = \frac{LD^2\pi}{4Q_0}, \qquad (4.41)$$

where D is the diameter of the mill and Q_0 is the volumetric velocity of the air flow.

Calculate now the gain factor m(t)/f(t)

$$N_1 = \frac{m(t)}{f(t)} = \frac{1}{\tau_0}\,\frac{1^T Uf'(t)}{f(t)} = \frac{1}{\tau_0}\,1^T Uf(t) = 1^T N_1 f(t) \qquad (4.42)$$

which is steadily changing, of course.

On the basis of the above, the slightly modified version of Fig. 4-4, better describing the real circumstances, is shown in Fig. 4-7.

The time constant of the closed-loop of material flow shown in the upper part of the Figure is

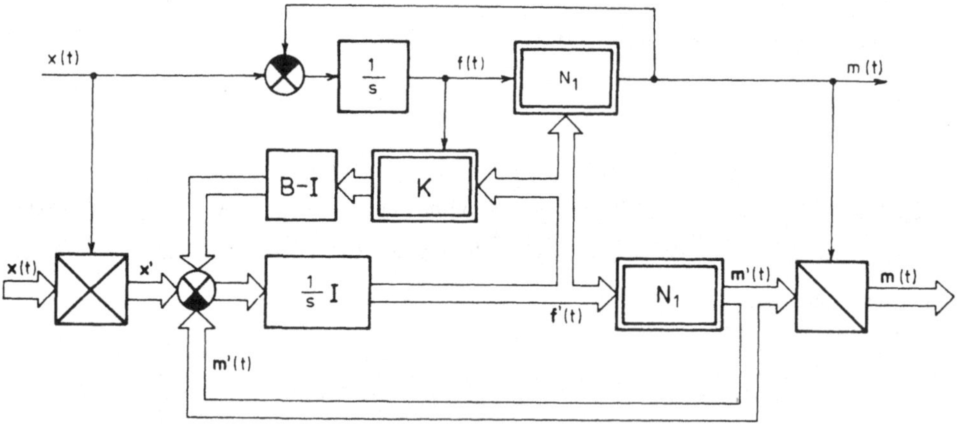

Fig. 4-7 Modified control engineering model for the quantitative and particle size distribution relations of the mill

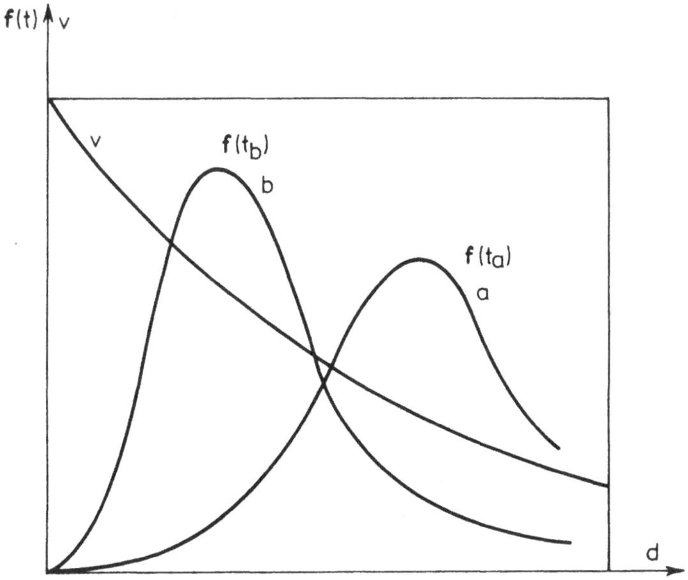

Fig. 4-8 The change in the particle size distribution of the material in the mill and the relative speed with respect to the fractions

$$T_1 = \frac{1}{N_1} = \frac{1}{1^T N_1 f} = \frac{\tau_0}{1^T \upsilon f} = \frac{\tau_0}{\sum_{i=1}^{n} \upsilon_i f_i} \tag{4.43}$$

(see more details in Section 6.1.)

Consider now Fig. 4-8 and assume that the size distribution (f(t)) of the material in the mill is changing from *a* to *b*, i.e. it is becoming finer. The Figure shows also the dependence of the relative velocity on the fractions, so it is obvious that the above change increases the denominator of (4.43), i.e. the time delay T_1 decreases. This feature of the model is in good agreement with the physical reality.

Let us summarize now the equations of our model, obtained as a final result by using the relation

$$m'(t) = N_1 f'(t) \tag{4.44}$$

and substituting it into (4.20) (thus (4.44) is valid now instead of (4.18)). We get

$$\frac{d\mathbf{f}(t)}{dt} = \{(\mathbf{B}-\mathbf{I})\mathbf{K}-[\frac{x(t)}{f(t)} - N_1]\mathbf{I}-\mathbf{N}_1\}\ \mathbf{f}(t) + \frac{x(t)}{f(t)}\mathbf{x}(t) \qquad (4.45)$$

and

$$\frac{df(t)}{dt} = [x(t)-N_1 f(t)] , \qquad (4.46)$$

where

$$N_1 = \mathbf{1}^T\mathbf{N}_1\mathbf{f}(t) = N_1[\mathbf{f}(t)] \qquad (4.47)$$

is dependent on the state variable. x(t) and **x**(t) are regarded as the inputs of the model, the state variables are f(t) and **f**(t).

The steady-state equation (4.22) becomes

$$[(\mathbf{B}-\mathbf{I})\mathbf{K} + (\mathbf{x}\mathbf{1}^T-\mathbf{I})\mathbf{N}_1]\mathbf{f} = \mathbf{A}_1(\mathbf{x})\mathbf{f} = 0 . \qquad (4.48)$$

Accordingly, the eigenvectors of $\mathbf{A}_1(\mathbf{x})$ have to be determined, of which the solution is $\mathbf{f}^*$, all of whose components are positive. (It can be proved easily that only one of the eigenvectors can fulfil this condition.) Finally, the "length" of the vector is determined so as to fulfil also the condition $\mathbf{1}^T\mathbf{f}^*=1$.

<u>Summary of Chapter 4</u>

First the distributed parameter one-dimensional model of open circuit continuous grinding based on equations describing the convection and mixing is discussed. This model is - in spite of many neglected terms - very sophisticated and is suitable mainly for simulation purposes and not for control.

Then - using the results of Chapter 3 - the control engineering model of open circuit continuous grinding is developed, where the size distribution **f**(t) and the material f(t) in the mill are regarded as state variables.

The results obtained are in good agreement with the results obtained from the **BROADBENT-CALCOTT** matrix models.

Finally, a modified dynamic model of the mill is derived, that is a better fit to reality. This model applies the relative fraction velocity and gives the interpretation of the outlet separation matrix. The inputs are x(t) and **x**(t), the state variables are f(t) and **f**(t).

Symbol nomenclature in Chapter 4

x(t)	feed [t/h]
m(t)	mill product [t/h]
x	discrete distribution approximating to the density function of the particle size distribution of feed
m	discrete distribution approximating to the density function of the particle size distribution of the mill product
$\mathbf{1}=[1,1,...,1]^T$	summing vector
f(t)	material in the mill [t]
f	density function belonging to the size distribution of the material in the mill
L	length of the mill [m]
l	axial distance in the mill [m]
E	coefficient matrix of second order dispersion
q	mass flow in axial direction [t/h]
f(t,l)	mass per unit axial length [t/m]
V	outlet separation matrix; time and space independent fraction velocity matrix
$\tau(t,l)$	mean residence time depending on space
v_i	relative velocity of particles in fraction i (relative to a mean velocity or velocity of the transport media)
f'(t),**m**'(t), etc.	partial material flows by fractions
N_1	nonlinearity, relationship between the mill product m(t) and the material f(t) in the mill
$\mathbf{N}_1$	outlet separation matrix
τ_i	residence time of the i-th fraction
v_i^*	corrected relative velocity of particles in fraction i
V_0	velocity of the transporting air flow

τ_0	residence time of the air
D	diameter of the mill
Q_0	volumetric velocity of the transporting air flow
$\mathbf{A}_1(\mathbf{x})$	coefficient matrix
$\mathbf{f}^*$	eigenvector having a physical meaning, i.e. the steady-state distribution vector

5. MATERIAL FLOW MODELS OF CLOSED-CIRCUIT GRINDING

The macrostructural models describing the material flows in closed circuit ball mill grinding include the rules of mass flow for steady-state, the nonlinear character of the mill, and the dead time relating to the delays in the system.

First, the static characteristics of closed circuit grinding in end-discharge mills are discussed in detail, then the influence of the particular nonlinearities on the dynamics of the closed-loop system is treated. The results obtained are then generalized to apply to central-discharge mills.

During the application of these lumped parameter models, the need arose to develop models that would better describe the transients of material flows due to change of grindability. The modified version of the material flow model given above can be considered as the first step in this respect. The dynamic size distribution models provide the final solution. Using the basic relationships of open-circuit continuous grinding, the dynamic size distribution model of closed-circuit grinding is also derived.

5.1. Macrostructural models

Let us consider the scheme of the end-discharge ball mill - classifier system in Fig. 5.1-1, where r(t), R is the fresh feed; g(t), G is the recirculating material (reject), x(t), X is the total feed; m(t), M is the mill product and v(t), V is the final product in tonnes per hour and c(t), C gives the revolutions per minute of the classifier. The lower-case letters refer to the time functions, the capital letters to their steady-state values, e.g. $m(\infty)=M$.

In this section the so-called macrostructural models of closed-circuit grinding suitable for characterizing the relationships between the material flows are investigated. These models should fulfil the rules of indestructibility of matter for steady-state operation, i.e. the following equations exist [115]

$$U = R \qquad (5.1.1)$$

$$M = X = R + G = U + G \,. \qquad (5.1.2)$$

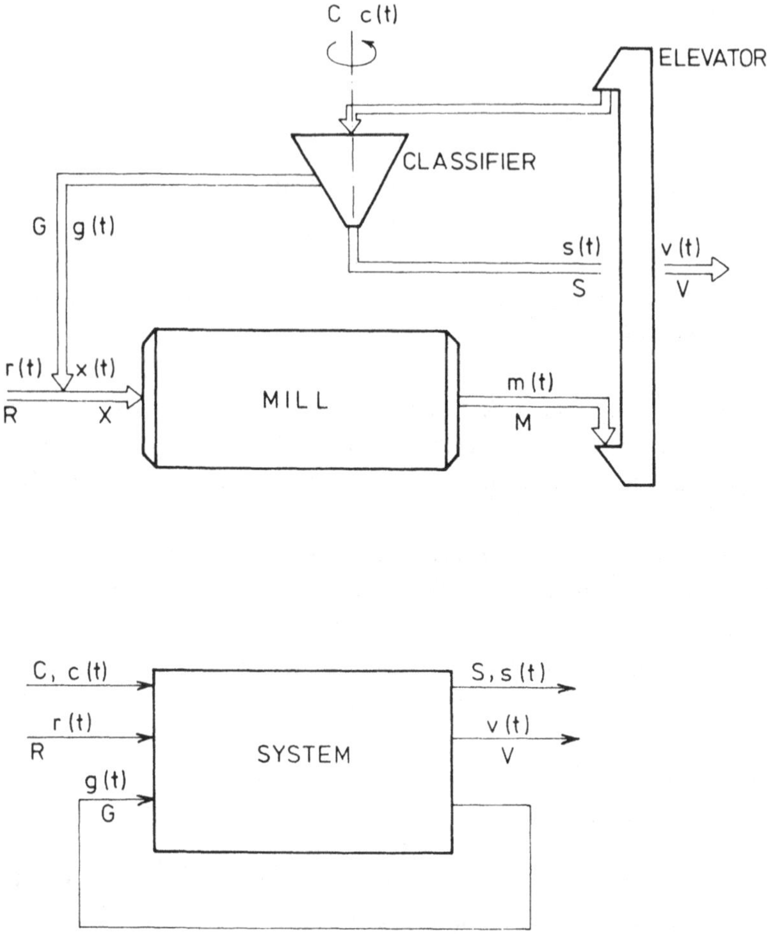

Fig. 5.1-1 Block-scheme of the end-discharge ball mill - classifier system

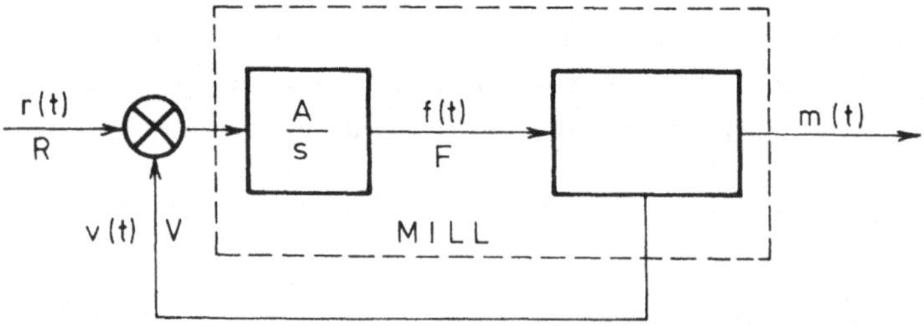

Fig. 5.1-2 Model for consideration of material filling factor

Considering the static behaviour of the mill, it has some special extremum properties. This means that the volume of product is limited. This limit to productivity represents an unstable point in the function of fresh feed and means an extremum (maximum) point as a function of the total feed or the volume of material in the mill.

The first quantitative models [202] introduced a measure for describing the ratio between the volume of material in the mill and the interstitial volume of the static ball charge, called the degree of filling (or material filling factor) f(t), F; see Fig. 5.1-2. Simultaneously, this Figure illustrates the rule of indestructibility of matter for the whole closed-circuit grinding system. The material f(t) in the mill is obtained as the integral of the difference of feed r(t) and outlet v(t) flows:

$$f(t)=A\int_0^t [r(t)-v(t)]\,dt \quad . \tag{5.1.3}$$

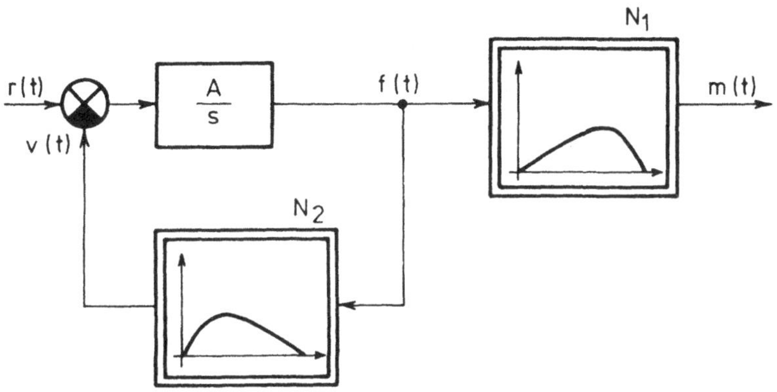

Fig. 5.1-3 Nonlinear mill model

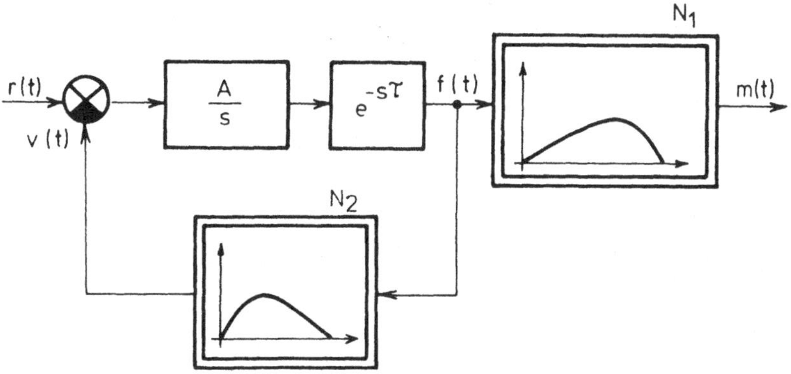

Fig. 5.1-4 Mill model with consideration of time delay

which corresponds to transfer function A/s. (Here A can be taken for unity in many cases; f(t) has to be measured in tonnes.) Note again that (5.1.2) is related to the input-output material flows of the whole closed-circuit grinding system, and in the steady-state (5.1.1) must hold true.

A more detailed version of Fig. 5.1-2 is shown in Fig. 5.1-3 where the nonlinearities N_1 and N_2 reflect the fundamental character of the mill. Fig. 5.1-4 also contains the dead time corresponding to the delay of the mill.

For dry mills (air-swept mills) the dead time τ is relatively small, but for tumbling mills it may be much longer. In practice, its values lie between the two cases. Furthermore it should be noted that - in the majority of cases in practice, in particular in the usual operating region - the nonlinearity N_1 can be considered to be linear with a good approximation. This means that the mill product, i.e. the load of the classifier, is proportional to the filling of the mill.

The most important phenomena are well explained by this model, and extremum control has already been performed [204]. It can be observed that if - according to the above - the mill product m(t) is actually proportional to the degree of filling f(t) then, by measuring m(t) and v(t), the derivative of the extremal characteristics N_2 can be obtained.

The measurement of the mill product m(t) is usually possible only by means of the power consumption of the elevator motor (this measurement is rather inaccurate with stochastic disturbances). There is no need for calibration, because if this power p(t), P is proportional to m(t), M, then the derivative dv(t)/dp(t) becomes zero at dv(t)/dm(t)=dv(t)/df(t)=0, which is the necessary condition of the extremum. Obviously, the final product has to be measured; furthermore, it is desirable to use stochastic filtering for p(t). To determine the position of the extremum, correlation and simple regression methods have been suggested; of course, advanced discrete identification procedures could be also applied [50],[118]. However, this approach and these methods are for use exclusively when the final product flow is measured continuously. This is, however, a not very general case at all; therefore, other approaches have to be sought.

The model according to Fig. 5.1-2 provides the fundamental mass balance R=V; however, in practice the mass balance M=X is of vital importance and R=V is obtained. Consequently,it is more desirable to establish the control system between x(t) and m(t) [115].

The model shown in Fig. 5.1-5 is far more complex than that of Fig. 5.1-4, which is developed on the basis of mass balance not valid for the whole grinding system but only for the mill, i.e. using

$$f(t) = \int_0^t [x(t)-m(t)]\,dt \quad . \qquad (5.1.4)$$

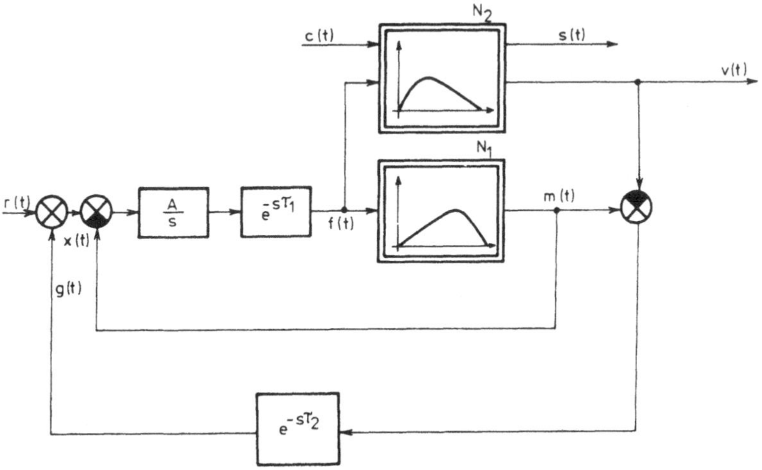

Fig. 5.1-5 General material flow model of the mill - classifier joint system

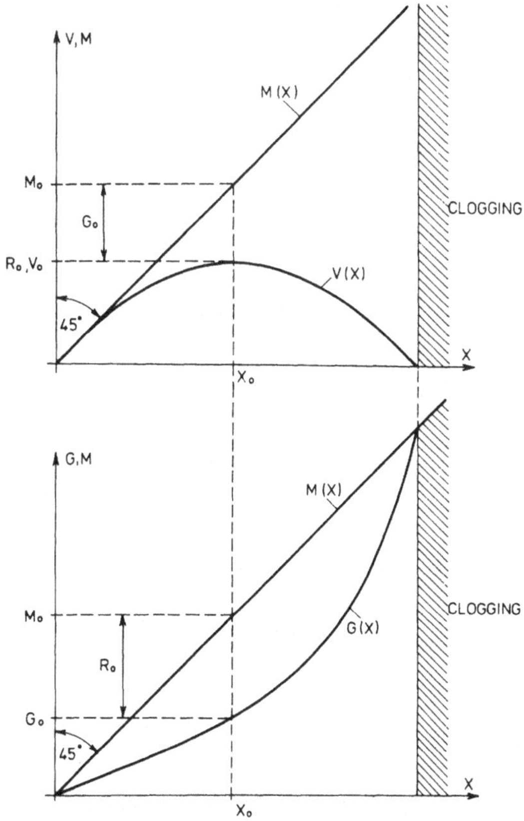

Fig. 5.1-6 Static characteristics of closed circuit grinding with respect to the total mill feed X

While in Fig. 5.1-4 f(t) means the material in the whole grinding system, in this case f(t) is related to the material in the mill. Thus now (5.1.2) means the basic relationship, since (5.1.4) gives also this in steady-state. The structure of the model represents well the real material flow relations and - as a consequence - yields (5.1.1). This model describes well, even down to detail, the quantitative behaviour of the mill and, disregarding the delays, it can be considered equivalent to Fig. 5.1-3 in respect of the final product.

Here the model contains only the main delays, but it is easy to include additional dead-time lags in all circuits because the geometrical arrangement is followed at the establishment of the control system.

In the inner loop τ_1 represents the dead-time of the mill. In the feed-back, τ_2 represents the delay of the conveyor of recirculated material. The block-scheme contains, in addition, the revolutions per minute of the classifier (separator) spreading plate c(t) and the specific surface of the final product, but they are disregarded for the moment and will be discussed later.

Let us summarize briefly the most important statements made on the static characteristics of closed-circuit grinding on the basis of Fig. 5.1-5. Figure 5.1-6 shows the functions M(X), V(X) and G(X) versus X. Due to the indestructibility of matter, M(X) is a straight line going through the origin with slope 45°, while V(X) has extremum at the optimal load X_0. For higher load, V(X) becomes zero, i.e. the mill gets clogged. Obviously, because of instability problems, the characteristics of the right-hand side are not desirable at all. The aim for optimal control is to seek, under varying circumstances, for the optimal load X_0, ensuring the optimal values of M_0 and V_0. (Here the varying circumstances may refer to the change of grindability and wear of the balls.)

Figs 5.1-7 and 5.1-8 give the static characteristics versus fresh feed R and degree of filling F, respectively.

Let us consider some typical examples to illustrate the properties of the above approach.

First, choose V(F) (i.e. N_2) as the function

$$V = 0.35\ F\ (40 - F)$$

which yields the maximum load V_{opt}=140 [t/h] at hold-up weight of the mill F_{opt}=20 [t]. Let N_1 be given by

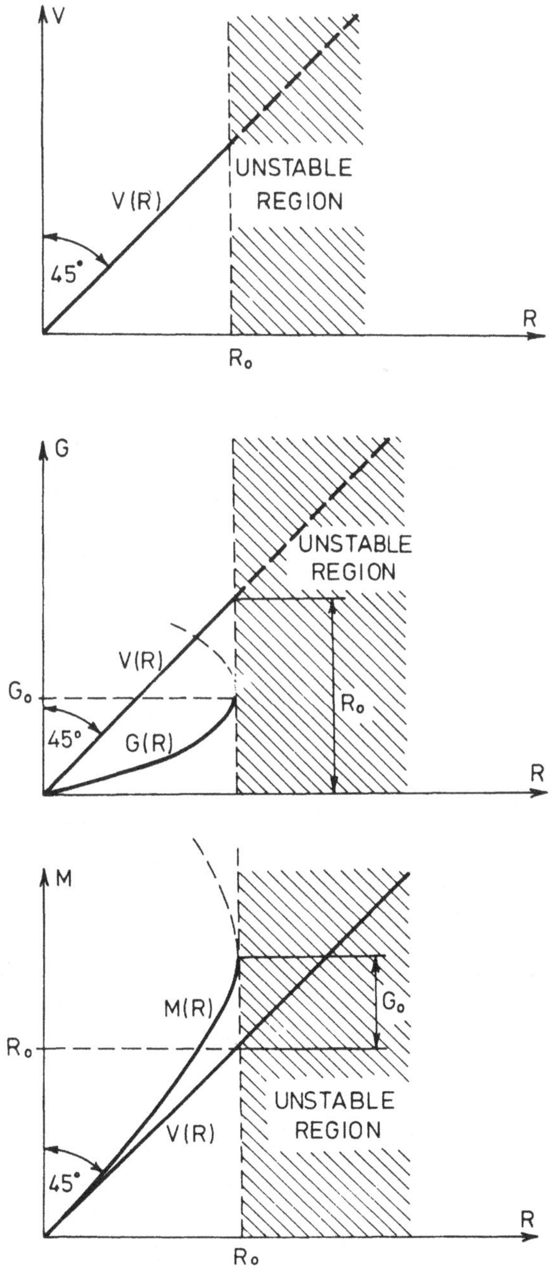

Fig. 5.1-7 Static curves with respect to the fresh feed R

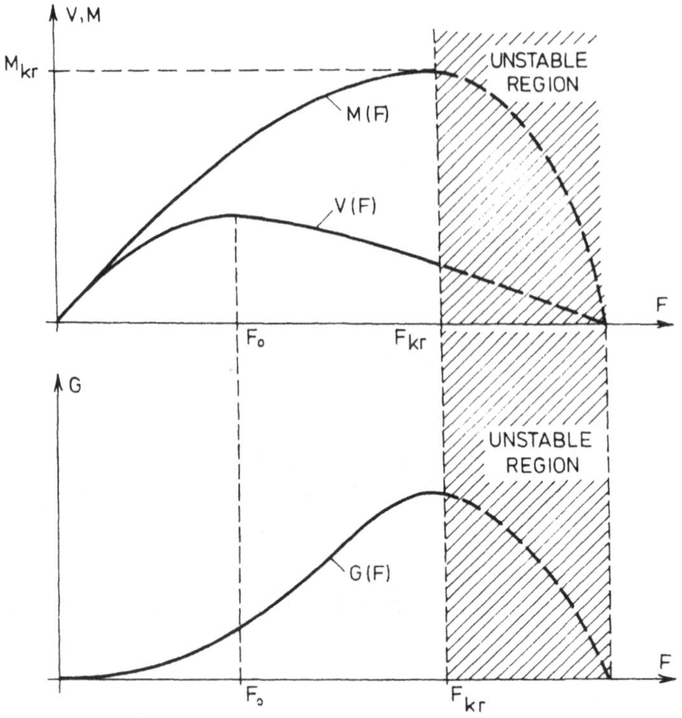

Fig. 5.1-8 Static curves with respect to the filling factor F

$$M = 0.25\,F\,(80 - F)$$

i.e. it is nearly linear over the region investigated. From the above equations the recycling material (grit) is

$$G = 0.1\,F\,(60 + F)$$

and the recycling rate [34] is

$$K = \frac{M}{U} = 0.714\,\frac{80-F}{40-F} \quad .$$

The curves V(F), M(F), G(F) and K(F) are drawn in Fig. 5.1-9; M(X), V(X), G(X) and K(X) are

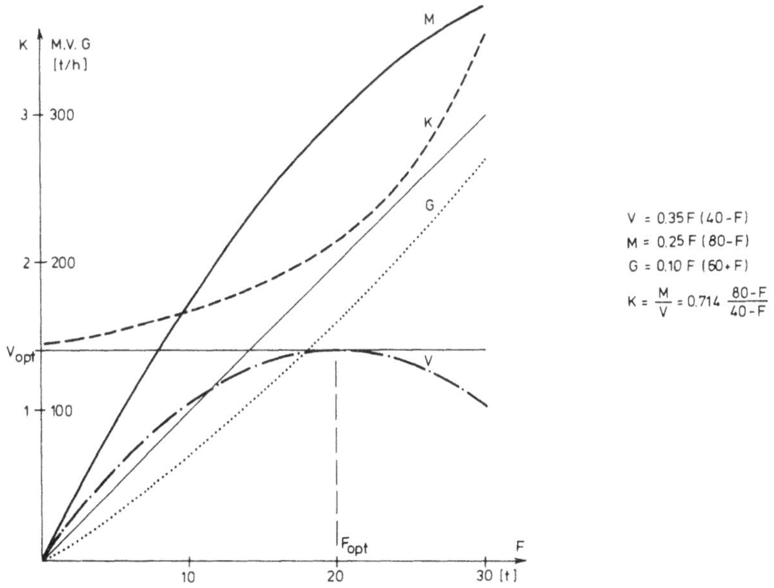

Fig. 5.1-9 Static curves of the closed circuit cement mill with respect to the filling factor

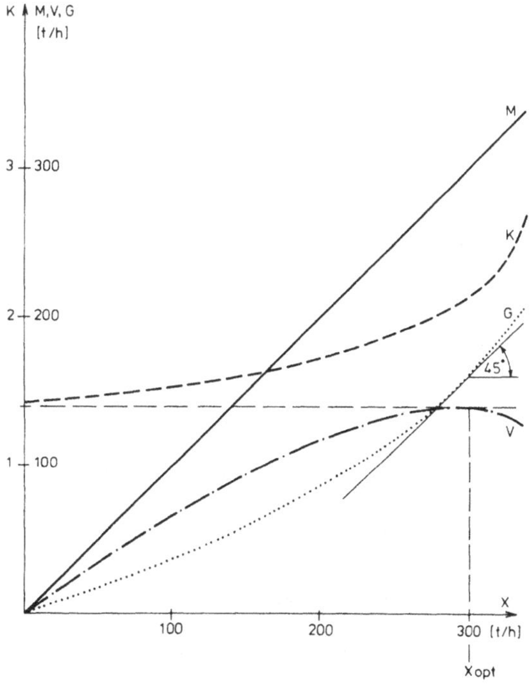

Fig. 5.1-10 Static curves of the closed circuit cement mill with respect to the total mill feed

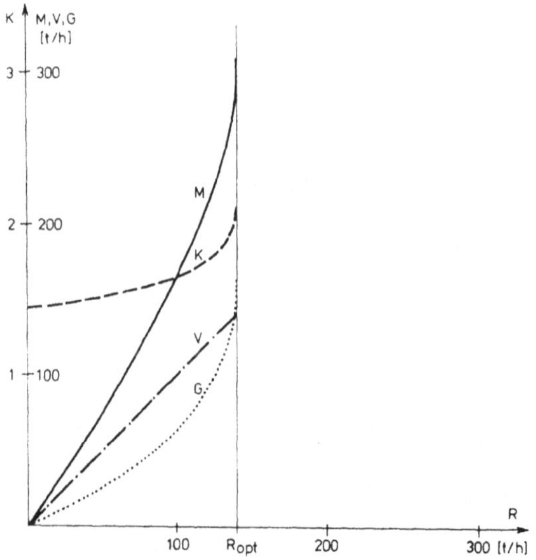

Fig. 5.1-11 Static curves of the closed circuit cement mill with respect to the fresh feed

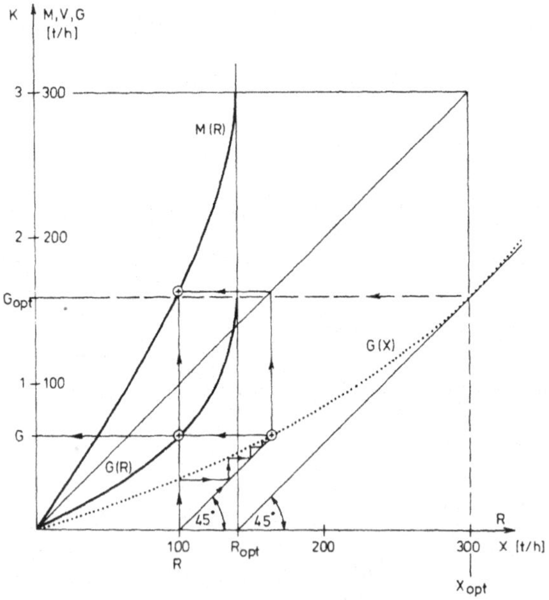

Fig. 5.1-12 Static curves of the closed circuit cement mill

in Fig. 5.1-10. The optimal load is at X_{opt}=300 [t/h] in accordance with practice. Similarly interesting shapes of curves are obtained for M(R), G(R), V(R) and K(R),shown in Fig. 5.1-11.

The properties considered in establishing the material flow model of the mill can be studied from these three Figures. The most important properties are listed below.

- The throughput V of the mill has a maximum value versus the degree of filling F. (This degree of filling appriximates to the quantity of material in the mill which fills the interstitial volume of the static ball charge,and that is when grinding is most effective. As the particle size distribution of the material is not uniform, the above property can only be interpreted by fractions, so it should be considered as an average behaviour.)
- The throughput V of the mill versus the total feed has the same maximum as the above. At this optimum working point the slope of G(X) is unity.
- There is a region for all material flows versus fresh feed R, where, over R_{opt}, material flows have no physical existence, so an optimal control needs the accurate assessment of the approach of an unstable boundary region. It should be mentioned that the extremum of N_2 occurs at a lesser degree of filling than the extrema of N_1.

  It is worth drawing all the curves in a common coordinate system. An example of this is shown in Fig. 5.1-12, which is an amended version of the diagram given in [34]. R and X are on the horizontal axis, and functions M(R), G(R), G(X) are plotted. It is seen that the values of G and M belonging to a given R can be obtained by the intersection of a line starting from R with a slope of 45°. Consequently, R_{opt} and X_{opt} can be drawn in the same way.

Here the line of slope of 45° is tangential to G(X). The existence of an optimum can be inferred from the fact that the curve G(X) has a portion where its tangent is greater than unity. (Because this is so, a straight line of slope of 45° must exist tangential to it somewhere!)

Fig. 5.1-13 shows figures in four plane sections illustrating how to draw the working point for a given feed R.

Now let us consider the case when the extremum point of nonlinearity N_1 appears at lesser degree of filling than the extrema of N_2.

Let

```
U = 0.35 F (40 - F) ;
```

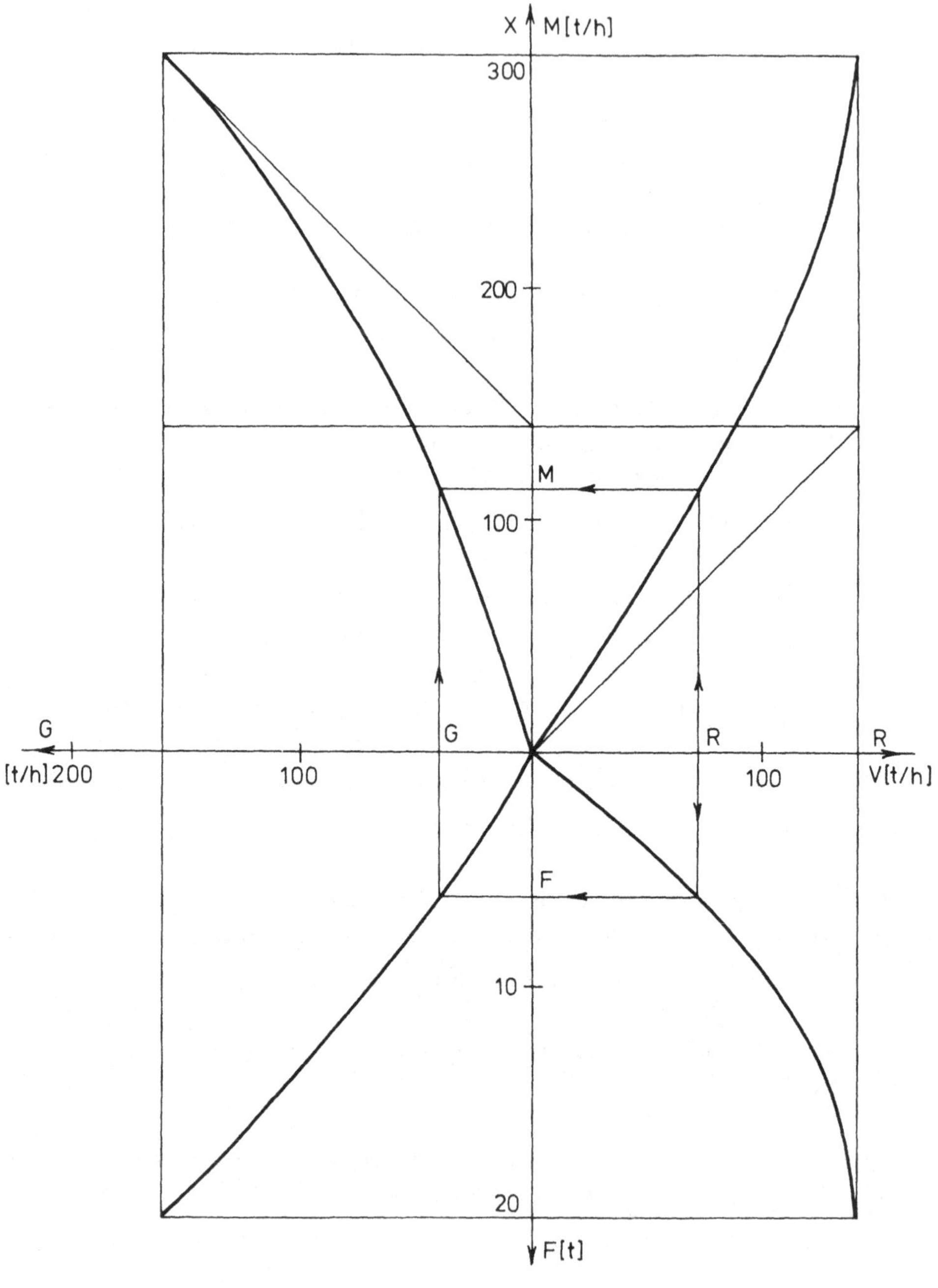

Fig. 5.1-13 Static curves of the closed circuit cement mill in four plane sections

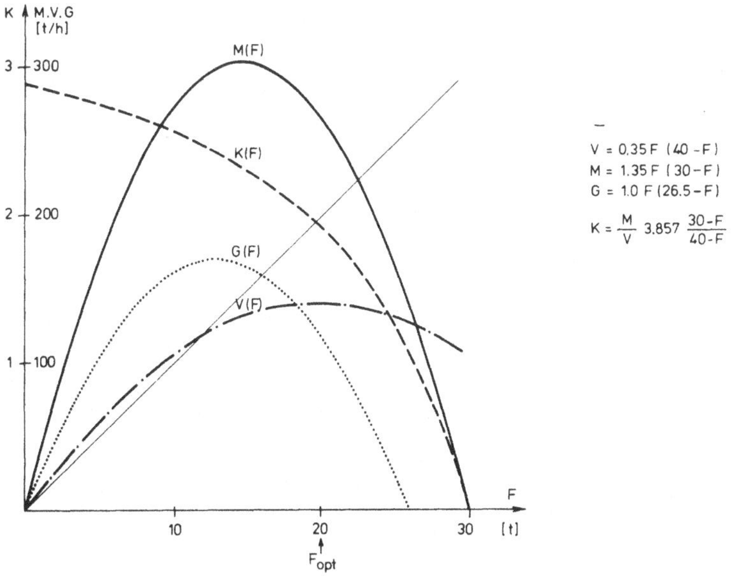

Fig. 5.1-14 Static curves for the case $F_1(N_{1max})<F_2(N_{2max})$ with respect to the filling factor

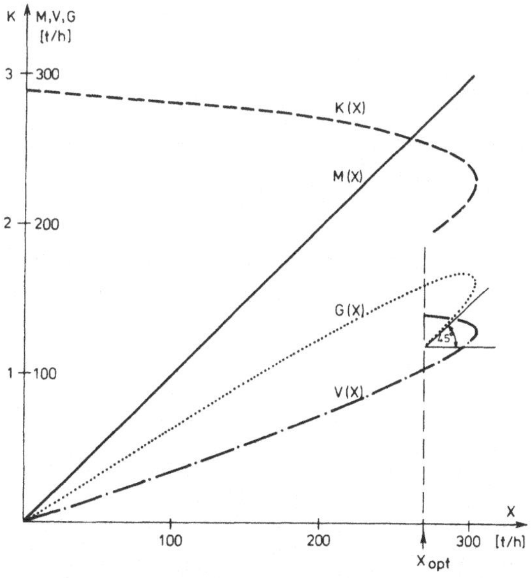

Fig. 5.1-15 Static curves for the case $F_1(N_{1max})<F_2(N_{2max})$ with respect to the total mill feed

furthermore,

$$M = 1.35\ F\ (30 - F)\ .$$

Hence

$$G = 1.0\ F\ (26.5 - F)$$

and

$$K = \frac{M}{U} = 3.857\ \frac{30-F}{40-F}\ .$$

The particular curves are presented in Fig. 5.1-14. The extrema of M(F) and G(F) occur at degrees of filling less than F_{opt}=20 [t], yielding the maximum of V.

The characteristics of M(X), G(X), V(X) and K(X) are shown in Fig. 5.1-15. Notice the reclinating character of the curves and the fact that the bivalent (bifurcating) sections also represent stable operational points. The rule in connection with the straight line having a slope of **45°** - though in a special form - is still valid in X_{opt}, yielding the maximum product.

In Fig. 5.1-16 the characteristics M(R), G(R), V(R) and K(R) are plotted versus fresh feed R. Here the extremum character of M(R) and G(R) is noticeable.

It has to be mentioned as a curiosity that this fact clearly distinguishes the two basically different cases as now the recycling rate K is decreasing for higher material flows, while in the previous case it increased unambiguously.

This fact also proves that the previous case (Figs. 5.1-9 ... 5.1-13) may be considered likely. On the other hand, in the literature, N_1 is taken to be linear rather than nonlinear.

Let us investigate the simplified version (Fig. 5.1-17) of the general model of closed-circuit grinding in Fig. 5.1-5. For the sake of simplicity, the dead times are not shown.

Determine the linear model of closed-circuit mill valid for infinitesimal deviations by linearizing the nonlinearities N_1 and N_2 around the operational point (see Fig. 5.1-18). Then the nonlinearities are

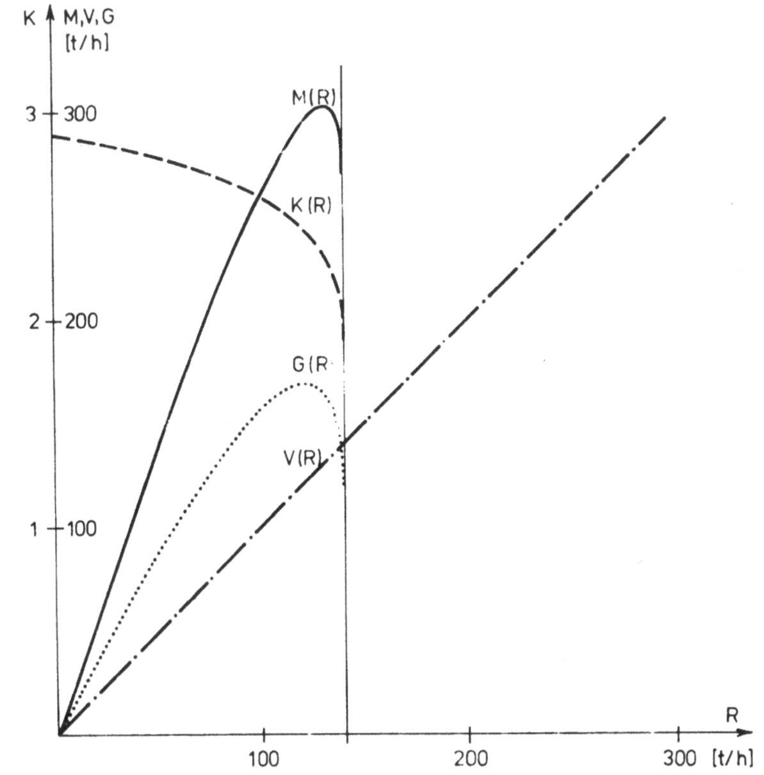

Fig 5.1-16 Static curves for the case $F_1(N_{1max})<F_2(N_{2max})$ with respect to the fresh feed

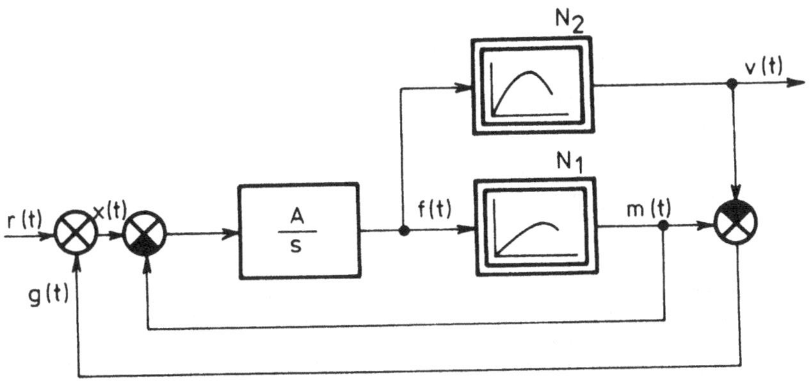

Fig. 5.1-17 Simplified version of the general model of closed circuit grinding

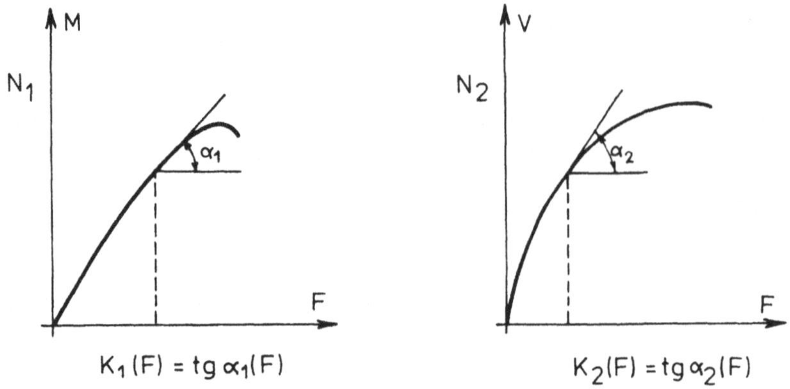

Fig. 5.1-18 Linearization of nonlinearities N_1, N_2 around the working point

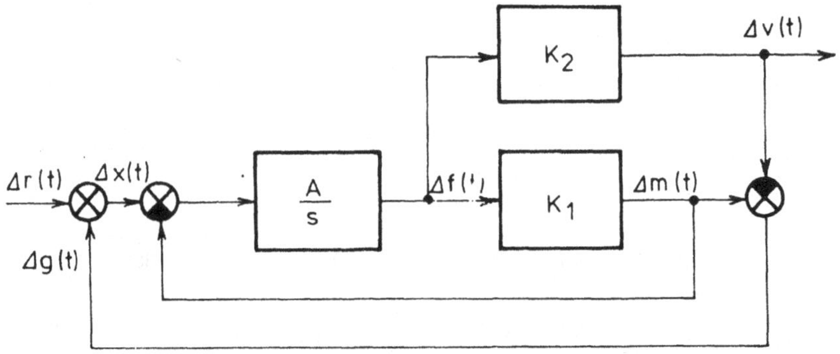

Fig. 5.1-19 Linear model of closed-circuit mill valid for small changes

taken into account by the following static functions:

$$K_1(F) = tg\ \alpha_1(F) = \frac{dM(F)}{dF}$$

and

$$K_2(F) = tg\ \alpha_2(F) = \frac{dU(F)}{dF}$$

depending on the degree of filling F.

Note that these derivatives become zero at the extremum of the nonlinearities, or become negative for the decreasing, unstable sections. Thus we get the linear system shown in Fig. 5.1-19. (Do not forget that now $\Delta r(t)$, $\Delta g(t)$, $\Delta x(t)$, $\Delta m(t)$ and $\Delta v(t)$ refer to small changes!)

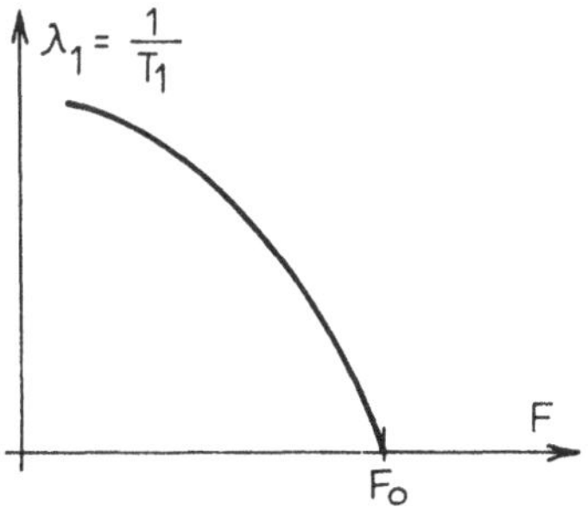

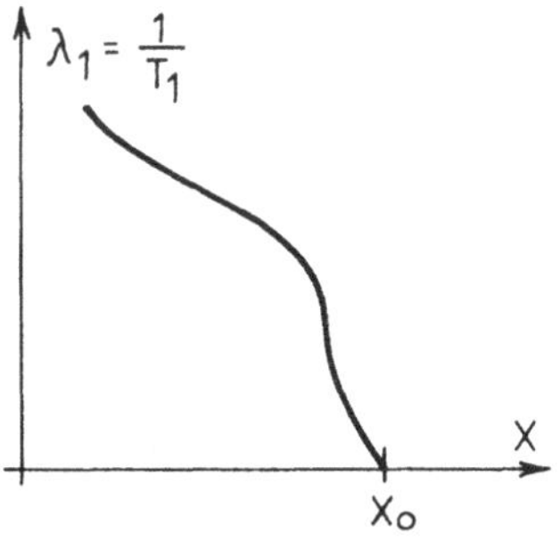

Fig. 5.1-20 The optimum filling factor and total mill feed values belonging to the nonlinearity N_1 dominant in closed system dynamics

Applying the rules of simplification of linear systems, the following transfer functions are obtained by Laplace transformation between the particular signals (changes):

1. When the input is the mill inlet (x(t)):

$$M(s) = \frac{1}{1+sT_1} X(s)$$

$$G(s) = \frac{K_1-K_2}{K_1} \frac{1}{1+sT_1} X(s)$$

$$U(s) = \frac{K_2}{K_1} \frac{1}{1+sT_1} X(s)$$

2. When the input is the fresh feed (r(t)):

$$M(s) = \frac{K_1}{K_2} \frac{1}{1+sT_2} R(s)$$

$$G(s) = \frac{K_1-K_2}{K_2} \frac{1}{1+sT_2} R(s)$$

$$U(s) = \frac{1}{1+sT_2} R(s) .$$

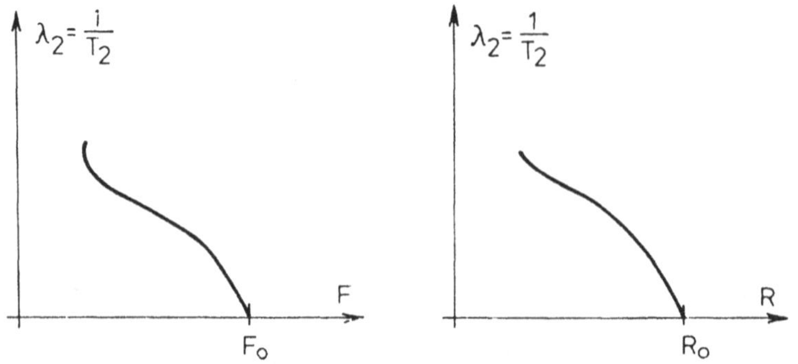

Fig. 5.1-21 Optimum filling factor and fresh feed values belonging to the nonlinearity N_2 dominant in closed system dynamics

Here

$$T_1 = T_1(F) = \frac{1}{AK_1(F)} \quad \text{and} \quad T_2 = T_2(F) = \frac{1}{AK_2(F)} \; .$$

The following conclusions can be drawn from the above relationships:

If the control signal is the mill inlet x(t) (total feed), then the dynamics of the closed system are determined by the nonlinearity N_1. In this case an input regulator is required, which adjusts the fresh feed r(t) according to the desired value of x(t).

At the extremum of N_1 the time constant T_1 becomes infinite. Accordingly, the quantity $\lambda_1=1/T_1=\lambda_1(F)$ becomes zero at the optimum degree of filling F_0. Obviously, the same phenomenon can be observed in connection with $\lambda_1(X)$; see Fig. 5.1-20. (Since, according to experience, the nonlinearity N_1 is considered to be nearly linear in the usual operation region of the mills, this phenomenon is not encountered in practice.)

If the control signal is the fresh feed r(t), then the behaviour of the closed system is determined by the nonlinearity N_2. As experience has shown, the optimization has to be directed to the determination of the extremum of N_2 rather than to the unstable region belonging to the extremum of N_1. At the extremum of N_2 the time constant T_2 becomes infinite, i.e. $\lambda_2=1/T_2=\lambda_2(F)$ becomes zero at the optimum degree of filling F_0. The same phenomenon can be observed in connection with $\lambda_2(R)$, see Fig. 5.1-21.

Although the functions $\lambda_1(X)$ and $\lambda_2(R)$ are nonlinear, they could be advantageously used for the estimation of the extremum, in particular in the vicinity of the optimum (and, of course, for the evaluation of the mill performance, considering the shapes of N_1 and N_2). Since it may happen that one single time constant does not provide adequately an accurate description of the real situation (think of the neglected dead-times), the extremum has to be sought by using the reciprocal value of the largest time constant of the transfer function.

In many cases r(t), g(t) and x(t) can be measured precisely, while m(t) can be estimated from the power p(t) consumed by the elevator. Under such conditions, the optimal working points can be estimated by identifying the relationships between the following signals:

When control is according to N_1, between

$$p(t) \; - \; x(t) \quad \text{and} \quad g(t) \; - \; x(t);$$

for control by N_2, between

$$p(t) \; - \; r(t) \quad \text{and} \quad g(t) \; - \; r(t) .$$

Up to now, material flow models of closed-circuit end-discharge ball mills have been dealt with. Let us investigate how to generalize these results for central-discharge mills. As we have done for end-discharge mills, only the most typical arrangement is discussed for central-discharge mills. We do not intend to go in detailed control engineering analysis of all possible grinding schemes.

Consider the closed-circuit, central-discharge ball mill arrangement shown in Fig. 5.1-22. The material flows are also shown in the Figure.

Essentially, the central-discharge mill can be regarded as two face-to-face connected end-discharge mills, having a special feature that $x_2=g_2$, i.e. there is no fresh feed on one (here on the right) side; only the grit is fed back. The generalized form of the end-discharge mill (Fig. 5.1-5) to a central-discharge one is shown in Fig. 5.1-23. Here and in the forthcoming, the equality

$$k_1 + k_2 = 1 \tag{5.1.5}$$

is assumed.

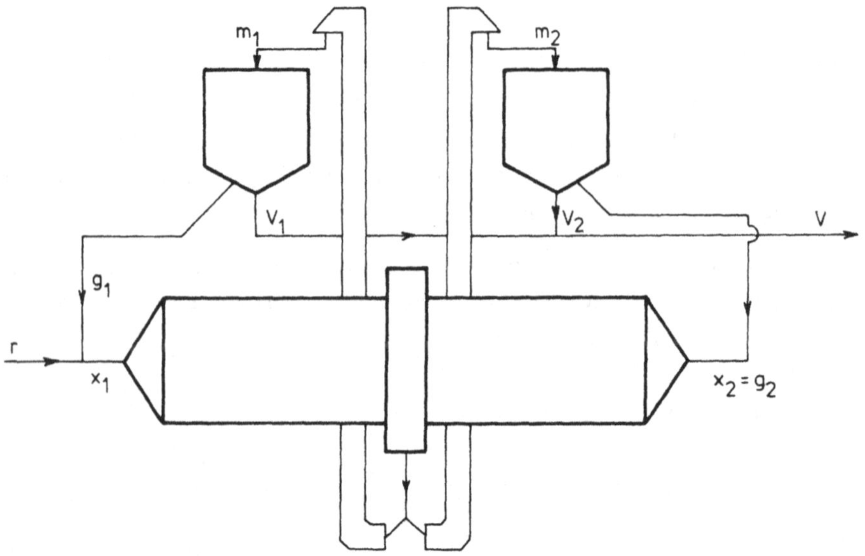

Fig. 5.1-22 Block-scheme of the closed-circuit central-discharge ball mill - classifier(s) joint system

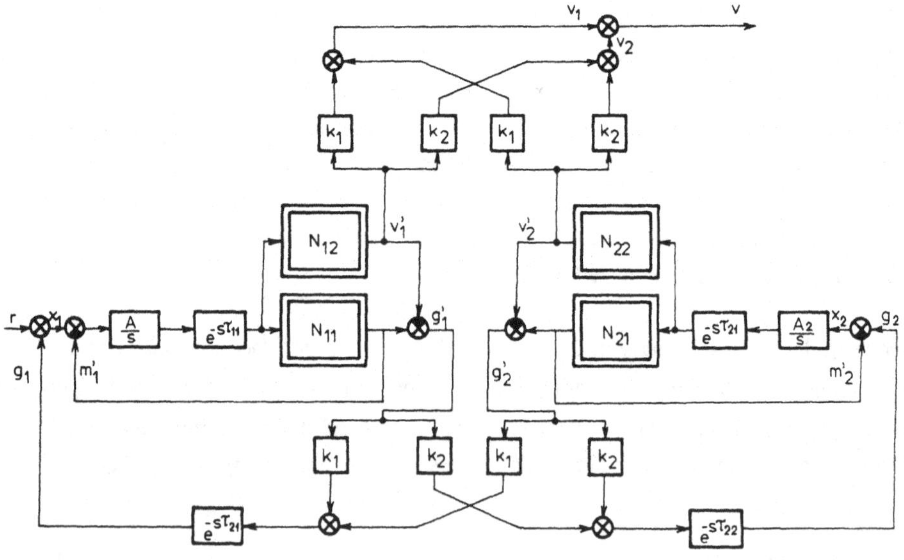

Fig. 5.1-23 Characteristic material flow of the closed-circuit central-discharge ball mill

For steady-state material flows (denoted by capital letters), the most important relationships of the central-discharge mill are:

$$X_1 = R + G_1 \tag{5.1.6}$$

$$X_2 = G_2 \tag{5.1.7}$$

$$G_1 = k_1 G_1' + k_1 G_2' = k_1 (G_1' + G_2') \tag{5.1.8}$$

$$G_2 = k_2 (G_1' + G_2') \tag{5.1.9}$$

thus

$$G_1 + G_2 = G_1' + G_2' \ . \tag{5.1.10}$$

Furthermore,

$$U_1 = k_1 (U_1' + U_2') \tag{5.1.11}$$

$$U_2 = k_2 (U_1' + U_2') \tag{5.1.12}$$

i.e.

$$U = U_1 + U_2 = (U_1' + U_2')(k_1 + k_2) = U_1' + U_2' \ . \tag{5.1.13}$$

To apply the model, the rate k_1/k_2 has to be known. It can be expressed as

$$\frac{k_1}{k_2} = \frac{U_1}{U_2} = \frac{G_1}{G_2} = \frac{M_1}{M_2} \ . \tag{5.1.14}$$

Fig.5.1-23 contains all the, sometimes fictitious, material flows needed to understand the operation of the system and to describe the mixing of these material flows. By neglecting the details of the inner structure, we get the simplified model shown in Fig. 5.1-24. Here the quantities v_i' and g_i' are omitted.

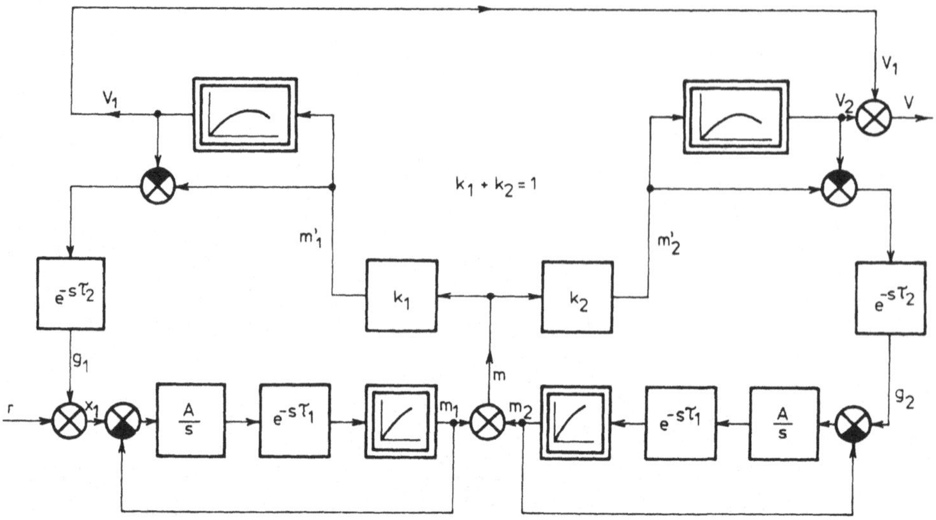

Fig. 5.1-24 Control engineering model of the closed-circuit central-discharge ball mill

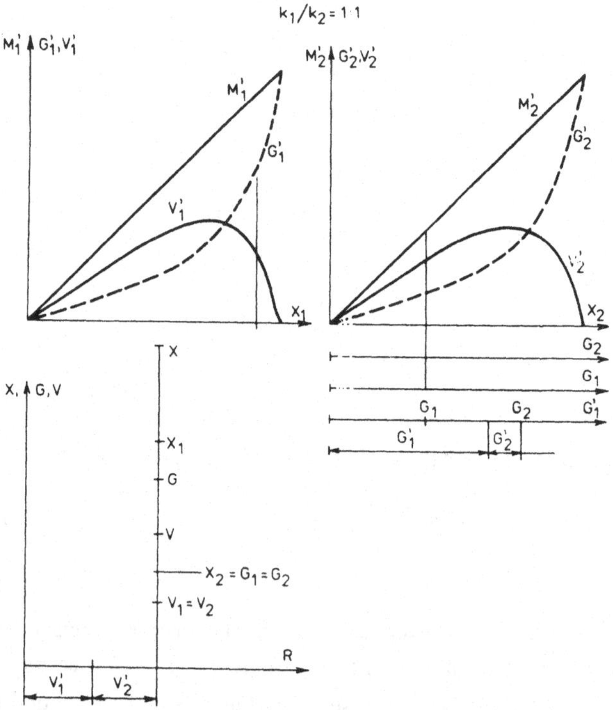

Fig. 5.1-25 Static curves of the closed-circuit central-discharge ball mill

Let us examine how to determine the resulting static characteristics of the whole mill from the particular characteristics of the two sides. Of course, the rate k_1/k_2 has to be known. The investigation is based on Fig. 5.1-25, where both sides of the mill are assumed to be the same and, for the sake of simplicity, $k_1/k_2=1$. In the Figure, the steps of the drawing can be followed singly, starting from the point $G_2=X_2$ on the second curve:

$$G_2 = X_2 \rightarrow U_2'$$
$$G_2'$$
$$M_2' = X_2$$
$$G_1 = \frac{k_1}{k_2} G_2 \rightarrow G = G_1 + G_2$$
$$G_1' = G_1 + G_2 - G_2'$$
$$U_1' \rightarrow U = U_1' + U_2'$$
$$M_1' = X_1 \rightarrow X_1 + X_2 = X$$
$$R = X_1 - G_1$$

Fig. 5.1-26 shows the material flows of both sides versus fresh feed R. Fig. 5.1-27 shows the characteristics of the central-discharge mill *observable* from outside which exactly correspond to those of the end-discharge mill. We call attention to the characteristic of $G_1(X_1)$ which at the optimum also shows as a straight line with slope 45°. This means that the optimal control of the central-discharge mill, with respect to the maximum product, can be also achieved by the measurement of the material flows of the left (coarse) side of the mill. (In general, however, it is assumed that - with an eventually incorrectly designed mill - the right side does not get clogged.)

Return again to the interpretation of the rule of straight line with slope 45°. For an end-discharge mill, one of the most important equations of indestructibility of matter is

$$M = U + G = X .$$

Differentiating by X, we get

$$\frac{dU}{dX} + \frac{dG}{dX} = \frac{dX}{dX} = 1 . \qquad (5.1.15)$$

It follows from the optimum condition

$$\frac{dU}{dX} = 0$$

that

$$\frac{dG}{dX} = 1 \qquad (5.1.16)$$

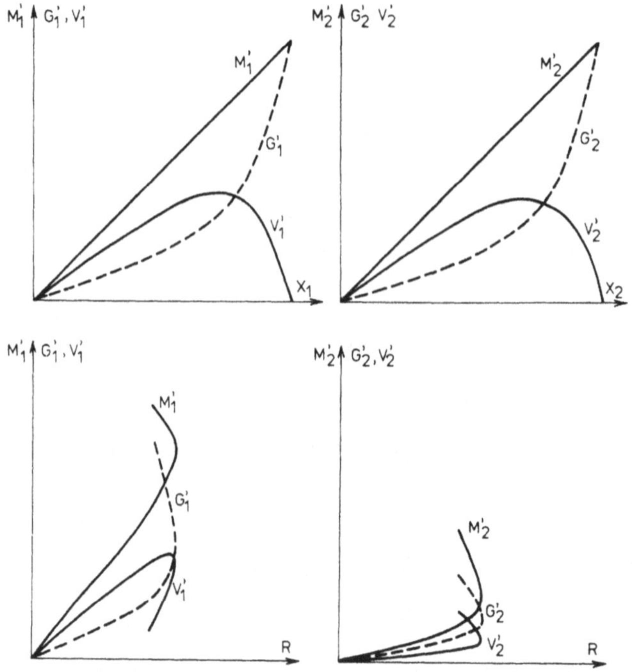

Fig. 5.1-26 *Inner* static curves of the closed-circuit central-discharge ball mill

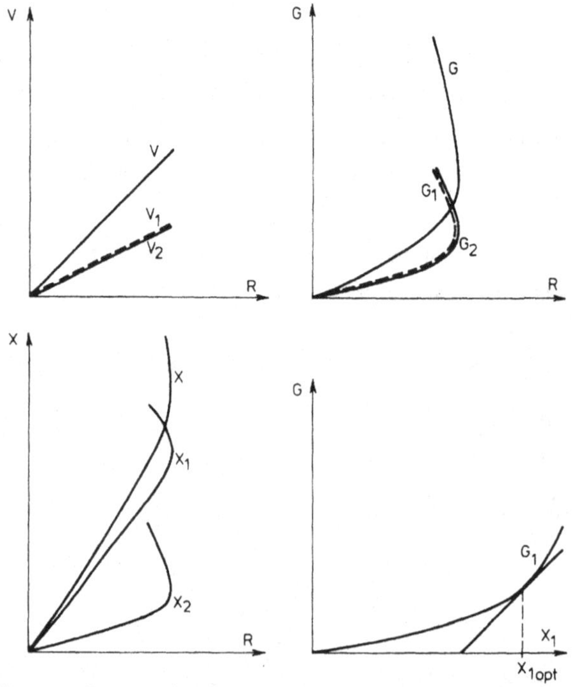

Fig. 5.1-27 *Outer* static curves of the closed-circuit central-discharge ball mill

which gives a straight line with slope $45°$.

For a central-discharge mill, the corresponding equation of indestructibility of matter is

$$X_1 = R + G_1 = U + G_1 \quad . \qquad (5.1.17)$$

Differentiating now by X_1

$$\frac{dX_1}{dX_1} = 1 = \frac{dU}{dX_1} + \frac{dG_1}{dX_1} \quad . \qquad (5.1.18)$$

To maximize the product the condition

$$\frac{dU}{dX_1} = 0$$

must hold true, i.e. again

$$\frac{dG_1}{dX_1} = 1$$

which also gives the above mentioned straight line.

Consequently, for closed-circuit grinding, the law of optimality is independent of the place of discharge and depends only on the place of feed.

It can be stated that the material flow models of closed-circuit grinding discussed above describe well the static characteristics of the end-discharge and central-discharge mills, i.e. their physical properties. These curves and model structures have made possible the explanation of most of the features of the grinding process, but not all. These lumped parameter models have provided great advances and new knowledge compared with earlier ones that were not system- and control engineering approaches. However, practical investigations have shown that these models need to be corrected in certain respects.

Theoretically, exact models can be developed by using particle size distributions, and this will be discussed in the next chapter. Since those models are already rather sophisticated, it seemed reasonable to apply certain generalizations with respect to the material flow models. As will be seen, the size distribution models certainly provide very accurate descriptions in all respects, using a lot of auxiliary quantities, while material flow models apply the integrated quantities measured in the plant by the instrumentation currently used in grinding mills, so they can be regarded as known variables.

Examining some recorded time functions of the material flows of end-discharge closed-circuit ball mills, it can be stated that there are oscillations in the time function of the mill product (or the power consumption of the elevator which is proportional to it), if the quantity of the grit or fresh feed is changing. This means that the input regulator used in order to keep the total feed on a constant level by changing the fresh feed with respect to the grit,is not suitable to prevent grit changes, resulting from the classifieradjustment or the qualitative requirements of the mill. The total feed remains constant, but meanwhile the composition or, more precisely, the recycling rate is changing. Actually the size distribution of the total feed is changing under constant material flow. Since in this case only integrated quantities are used, the changes have to be described by quantitative and qualitative measures of the total feed.

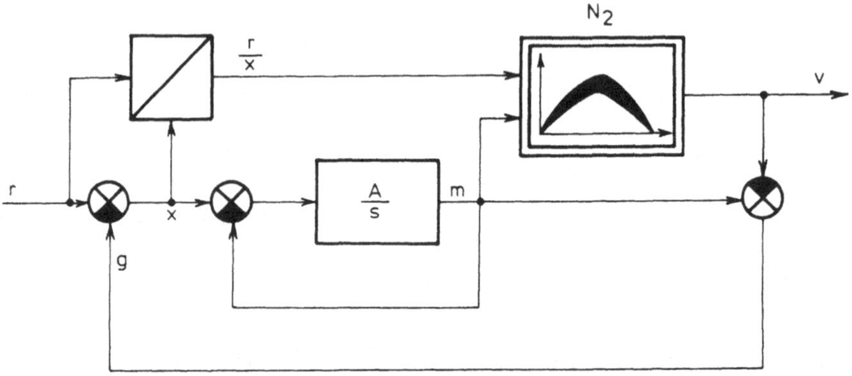

Fig. 5.1-28/a Control engineering model for taking the transient changes in nonlinearity N_2 into account

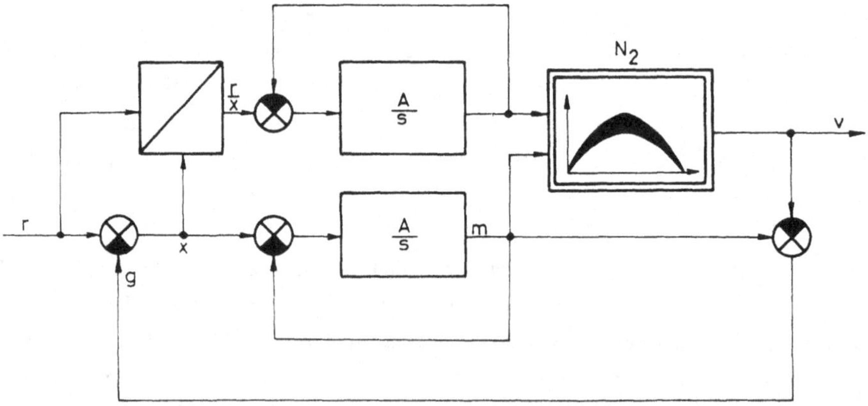

Fig. 5.1-28/b Modified control engineering model for taking the transient changes in nonlinearity N_2 into account

If the structure of the total feed is taken into account by the ratio g/x, two phenomena can be distinguished while the rate is changing from zero to one, and x is a constant. One is that the mill product, i.e. the power consumption of the elevator which is proportional to it, increases. Obviously, this may be only a transient, dynamic effect, because in the steady-state, the equality M=X must hold true. Essentially, the matter is due to the decrease of the residence time in the mill caused by the increased grit ratio, i.e. the mill is emptied faster, thus the gain of the inner control loop increases. (The time constant of the discharge, i.e. of the decrease of f(t) is $T_1=1/A_1$, where A_1 means the gain of the inner loop of the mill.) However, the classifier can separate more of the finer particles, so the nonlinearity N_2 can produce more final product; as a consequence, it becomes elongated along the vertical axis.

Fig. 5.1-28/a shows the block-scheme of a model which is based on existing results and knowledge; it describes the transient behaviour of the material flows of the mill in the above manner. The composition of x is represented by the ratio

$$\frac{1}{k} = \frac{r}{x} , \qquad (5.1.19)$$

i.e. by the reciprocal value of the recycling rate. For the nonlinearity N_2 the possible changing range of the static characteristic, yielding higher outlet ratio, is shown black in the Figure.

Because the ratio r/x propagates through the mill with approximately the same time constant, a same closed loop, representing a first order lag, should be applied in its signal flow path, which is already used in Fig. 5.1-28/a. This resulting block-scheme is shown in 5.1-28/b.

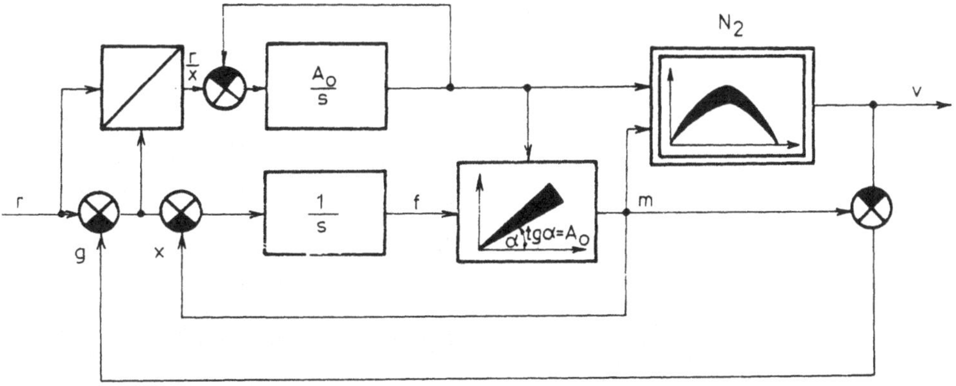

Fig. 5.1-29 Control engineering model taking the transients of the material flows of the mill into account

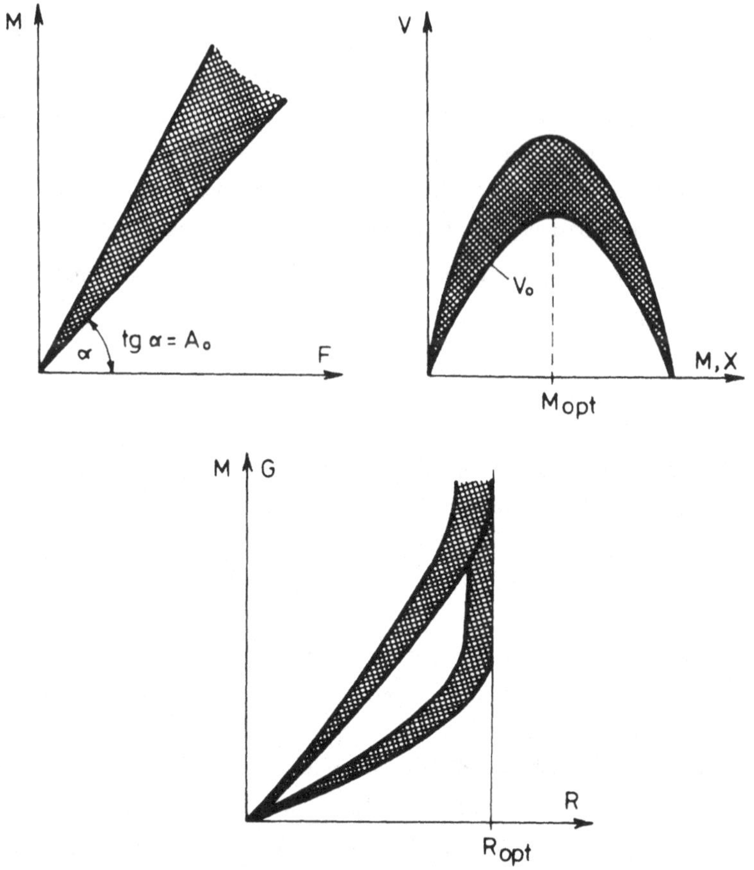

Fig. 5.1-30 Uncertainty region of the static curves caused by the unmodelled dynamics

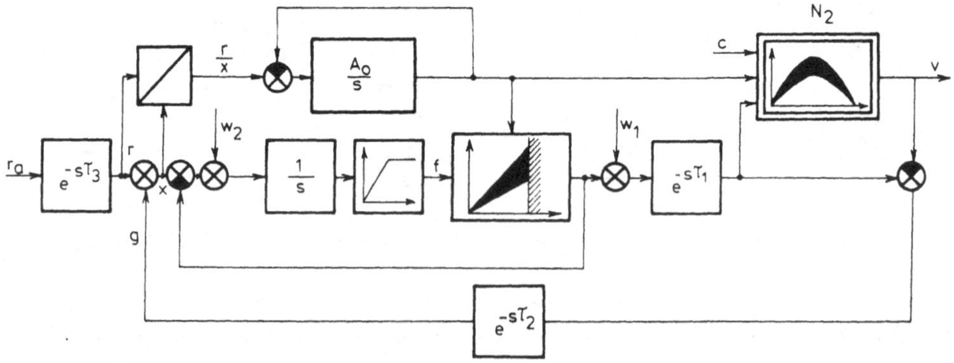

Fig. 5.1-31 Control engineering mill model completed by time delays and restrictions

The two kinds of phenomena can be even better followed in Fig. 5.1-29, where the changing range for the static gain of the inner block between f and m, yielding the decrease of the residence time mentioned above, is also drawn in black. (The gain of the integrator is chosen to be unity.) From the quantitative aspect, these relationships can be written in the following form. Let the nonlinearity N_2 be

$$V(R,X,M) = V(G,X,M) = V_0(M)[1+\beta_V(\frac{G}{X} - \frac{G_0}{X_0})] =$$

$$= V_0(M)\ [1-\beta_V(\frac{R}{X} - \frac{R_0}{X_0})] . \qquad (5.1.20)$$

Here $V_0(M)$ represents the original characteristics and β_V is an incremental factor relating to the ratio G/X (whose value is small; it depends on M and many other factors).

The relationship between M and F is

$$M(R,X,G)=M_0(F)[1+\beta_M(\frac{G}{X} - \frac{G_0}{X_0})]=M_0(F)[1-\beta_M(\frac{R}{X} - \frac{R_0}{X_0})]. \quad (5.1.21)$$

Here β_M is an incremental factor also relating to the ratio G/X. It has to be remarked that these new expressions have no considerable influence on the static characteristics of the mill. The extremal features have remained, however; their direct calculation is not a trivial matter.

Eqs. (5.1.20) and (5.1.21) contain capital letters (thus they refer to steady-state conditions), but they are also valid for time functions. It becomes obvious from the static equations valid for steady-state that they could be transformed further, giving the usual characteristics M(F), V(F), G(F), V(X), G(X), X(R), G(R), etc., in the form of explicit expressions which are already too complicated; therefore,it is more reasonable to apply numerical solutions.

At the same time the transient properties of the new model differ radically from the dynamic behaviour of the previous, in practice first order, model. Fig. 5.1-30 shows the situation when the static characteristics can be perceived only with an *uncertainty* region because of the steady fluctuation of the recycling rate due to an inadequate control strategy; consequently, the determination of the optimal working point becomes difficult. Thus, to retain maximum V(M), it is not enough to adjust X: the steady-state value of m(t) must be kept constant by the input regulator; and the recycling rate X/R has to be kept on a constant level.

In Fig. 5.1-31 the model according to Fig. 5.1-29 is completed with the inclusion of the most important time delays of the real system, i.e. with dead-time lags. Here τ_1, τ_2, τ_3 refer to the time delays of the bucket elevator and the conveyor belts carrying the grit and fresh feed, respectively. A saturation element is included in the inner loop representing an eventual saturated state of the first compartment of the mill. The quantities w_1 and w_2 are stochastic disturbances with zero mean, from which w_1 represents the stochastic fluctuations connected with the measurement and transport of the bucket elevator, while w_2 indicates the random changes in the grinding conditions, the grindability. The nonlinearity N_2 includes also the relationship with respect to the revolutions per minute of the classifier (c), which requires further discussion. According to the present state of our knowledge, of all material flow models of mills, this version can be considered as the most advanced.

On the basis of (5.1.20) and (5.1.21) the characteristics V(X) and G(X) can be determined in an explicit form. Taking the equation of indestructibility of matter X=R+G=V+G=M into account, and using (5.1.20), we get

$$U(X) = U_0(X)\left[1-\beta_U\left(\frac{U}{X} - \frac{U_0}{X_0}\right)\right] = U_0(X)\,\frac{1+\beta_U\dfrac{U_0}{X_0}}{1+\beta_U\dfrac{U_0}{X}} . \qquad (5.1.22)$$

5.2. Dynamic particle size distribution models

In every stage of grinding, in every section of the open-circuit mill, the material which is getting finer occurs in the form of a given size distribution, but - depending on the initial state and external mill variables - with changing parameters, i.e. it contains fine fractions in a rather considerable quantity. Closed-circuit grinding is based on the separation of the fine from the oversize fractions and the recirculation of the coarse fractions. The principle of closed-circuit grinding is shown in Fig. 5.2-1 and is in accordance with the content of section 5.1. In addition, as mentioned there, r(t), g(t) and v(t) are now the density vectors representing the discrete size distribution of fresh feed, grit and final product, respectively, containing the relative percentages by weight (or the probability of falling into a given fraction).

The equations of open-circuit grinding discussed in Chapter 4 have to be completed by the following equations applying the basic relations in Section 5.1.:

$$x(t)=r(t)+g(t)\ ; \qquad v(t)=m(t)-g(t) \tag{5.2.1}$$

$$x'(t)=r'(t)+g'(t) \tag{5.2.2}$$

$$x(t)x(t)=r(t)r(t)+g(t)g(t) \tag{5.2.3}$$

whence

$$x(t) = \frac{r(t)}{x(t)}\, r(t) + \frac{g(t)}{x(t)}\, g(t)\ . \tag{5.2.4}$$

The operation of the classifier can be described by a diagonal matrix **T** where the elements of the main diagonal represent the percentage by weight (or probability) of falling into the coarse fraction, i.e. into the grit, of that particular fraction.

(The computation of the elements of matrix **T** will be discussed in Chapter 6.) According to the above, the density function of the grit

$$g'(t) = g(t)g(t) = m(t)Tm(t) = T\ m'(t) \tag{5.2.5}$$

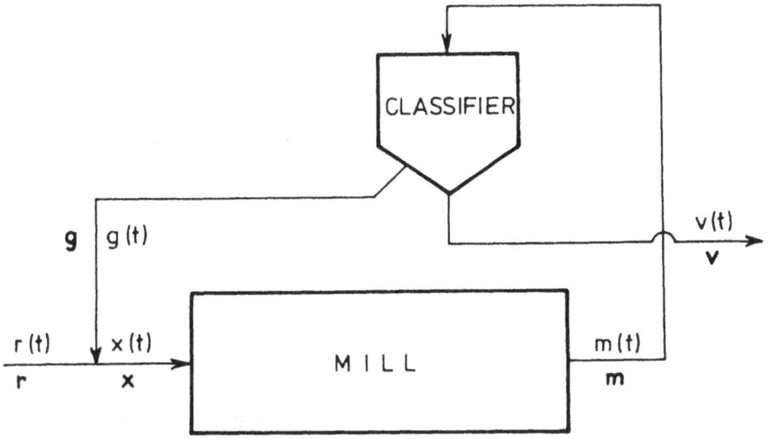

Fig. 5.2-1 Block-scheme of closed-circuit grinding including the particle size distribution

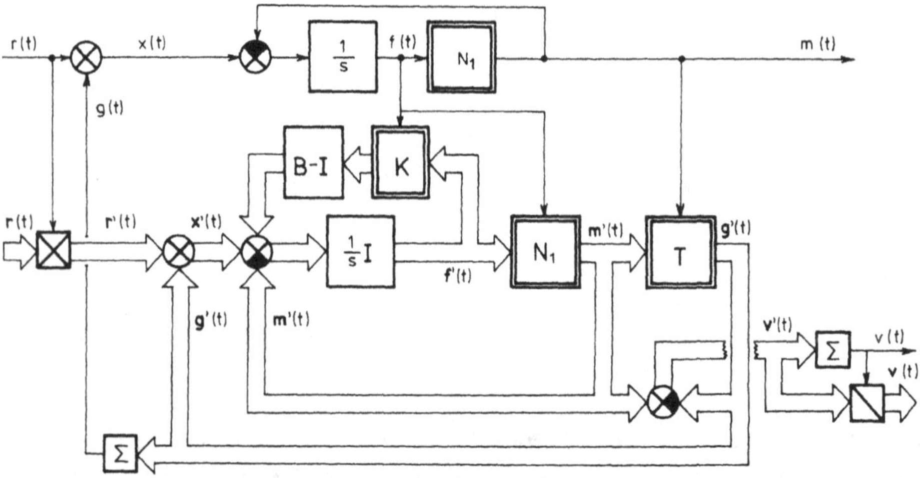

Fig. 5.2-2 Block-scheme of the grinding system taking quantitative and particle size distribution relations into account

can be determined from the mass balance in the screens (i.e. by fractions). Based on the equations written up to now, and on Fig. 4-7, a possible block-scheme of the system, shown in Fig. 5.2-2, can be established. (This scheme can also be extended by the transporting time delay of the bucket-elevator or by that of the grit conveyer belt, similarly to Fig. 5.1-5.)

In the model Σ refers to the multiplication by $\mathbf{1}^T$, by means of which the material flows corresponding to the integrated quantities are determined.

It can be well seen that the model contains all the components of closed-circuit grinding technology and is a very effective tool to describe the interaction between *quantitative maximization* and *optimal fineness.*

Using (5.2.4), (5.2.5) and (4.44), we can write

$$x(t) = \frac{r(t)}{x(t)} r(t) + \frac{1}{x(t)} T\, m'(t) = \frac{r(t)}{x(t)} r(t) + \frac{f(t)}{x(t)} T\, N_1 f(t) \tag{5.2.6}$$

Substituting (5.2.6) into (4.45), we get the differential equation of the state vector f(t) of closed-circuit grinding

$$\frac{d\mathbf{f}(t)}{dt} = \{(\mathbf{B}-\mathbf{I})\mathbf{K} - [\frac{x(t)}{f(t)} - N_1]\mathbf{I}+(\mathbf{T}-\mathbf{I})N_1\}\mathbf{f}(t) + \frac{r(t)}{f(t)}\mathbf{r}(t) =$$

$$= \{(\mathbf{B}-\mathbf{I})\mathbf{K} - [\frac{r(t)}{f(t)} - N_1]\mathbf{I} + (\mathbf{T}-\mathbf{I})N_1\}\ \mathbf{f}(t) + \frac{r(t)}{f(t)}\mathbf{r}(t) \tag{5.2.7}$$

as well as that based on (4.11),

$$\frac{df(t)}{dt} = r(t) + g(t) - m(t) = r(t) - v(t) = r(t) - N_2 f(t) \tag{5.2.8}$$

which constitute the dynamic size distribution model of closed- circuit grinding. The state variables are f(t) and **f**(t), but now the input variables are r(t) and **r**(t). In the last two equations the equality

$$v(t) = m(t) - g(t) \tag{5.2.9}$$

is assumed, and the state-dependent factor

$$N_2 = \frac{v(t)}{f(t)} = \frac{m(t)-g(t)}{f(t)} = \mathbf{1}^T(\mathbf{I}-\mathbf{T})N_1\mathbf{f}(t) = N_2[\mathbf{f}(t)] \tag{5.2.10}$$

is introduced. Eqs (4.44) and (5.2.5) have also been used.

The steady-state size distribution **f** is also of interest. Considering that in the steady-state d**f**(t)/dt=**0** and r(∞)=v(∞), then using (5.2.10), from (5.2.7)

$$[(\mathbf{B}-\mathbf{I})\mathbf{K} + (\mathbf{I}-\mathbf{r}\mathbf{1}^T)(\mathbf{T}-\mathbf{I})N_1]\mathbf{f} = \mathbf{A}_r(\mathbf{r})\mathbf{f} = \mathbf{0} \tag{5.2.11}$$

is obtained, whose solution is to be sought as in the case of (4.48). With the solution known, further density vectors can be obtained by simple calculation:

$$\mathbf{v} = \frac{(\mathbf{I}-\mathbf{T})N_1\mathbf{f}}{\mathbf{1}^T(\mathbf{I}-\mathbf{T})N_1\mathbf{f}} \quad ; \tag{5.2.12}$$

$$\mathbf{g} = \frac{\mathbf{T}N_1\mathbf{f}}{\mathbf{1}^T\mathbf{T}N_1\mathbf{f}} \quad ; \tag{5.2.13}$$

$$m = \frac{N_1 f}{1^T N_1 f} \; ; \tag{5.2.14}$$

$$x = \frac{1^T (I-T) N_1 f}{1^T N_1 f} r + \frac{T N_1 f}{1^T N_1 f} = r + \frac{(I - r 1^T)}{1^T N_1 f} f \tag{5.2.15}$$

Summary of Chapter 5

Macrostructural models of closed-circuit (ball mill) grinding, satisfying the laws of steady-state material flow,are determined, to include the nonlinear, extremum character of the mill and to contain the dead times corresponding to particular time delays.

After a detailed investigation of the static characteristics of closed-circuit end-discharge mills, the effect of the particular nonlinearities on the dynamics of the closed system is shown. Then the results are generalized for central-discharge mills.

During the practical application of these lumped parameter models the need arose to develop models describing better the transient nature of the material flows due to changes in grindability. The modified material flow model elaborated here can be considered as the first step in this direction.

The dynamic size distribution models lead to the final, most accurate solution. Applying the basic relationships of open-circuit continuous grinding, the dynamic size distribution model of closed-circuit grinding is derived.

Symbol nomenclature in Chapter 5

(The lower-case letters refer to the time functions, the capital letters stand for their steady-state values.)

r(t), R	fresh feed [t/h]
g(t), G	grit [t/h]
x(t), X	total feed [t/h]
m(t), M	mill product [t/h]
v(t), V	final product [t/h]
f(t), F	degree of filling [t]
p(t), P	power consumption of the bucket-elevator's motor drive [kW]

A	gain factor (here A=1 is assumed)
s	Laplace operator
N_1, N_2	scalar nonlinearities
τ_i	dead-time [min]
$c(t)$, C	revolutions per minute of the spreading plate in the classifier
K	recycling rate
k	instant value of the recycling rate
$K_1(F)$, $K_2(F)$	static gain factors
Δ	infinitesimal element
T_1, T_2	time constants [sec]
$\lambda_1()$, $\lambda_2()$	proper functions for the estimation of the extremum of the nonlinearities N_1, N_2
k_1/k_2	ratio between the material flows getting to the different compartments of the central-discharge mills
V_i', G_i', M_i'	inner material flows of the central-discharge mill
β_V, β_M	inceremental factors relating to static curves
w_i	stochastic disturbance with zero mean [t/h]
$\mathbf{r}(t),\mathbf{g}(t),\mathbf{x}(t),\mathbf{m}(t),\mathbf{v}(t)$, etc.	density vectors containing relative percentages by weight, representing the size distribution of particular material flows
$\mathbf{r}'(t),\mathbf{g}'(t)$,..., etc.	the absolute value vectors of partial material flows falling into corresponding fractions [t/h]
$\mathbf{T}$	diagonal matrix representing the operation of the classifier
Σ	summation (multiplication by $\mathbf{1}^T$) to form the corresponding integrated quantities
$\mathbf{I}$	unit matrix
$\mathbf{N}_1$	matrix of nonlinearity
$\mathbf{B}$	comminution matrix
$\mathbf{K}$	breakage rate matrix
$\mathbf{A}_r$	coefficient matrix

6. MATHEMATICAL MODEL OF THE CLASSIFIER

It has been seen that in the model of closed-circuit grinding the classifier is taken into account by a diagonal classification matrix **T** corresponding to the so-called **TROMP** curve. The main diagonal elements give the probability of the given fraction falling into the coarse product [222].

The **TROMP** curves of an ideal classifier and an actual one are presented in Fig. 6-1. A very specific parameter of the classifier is d_{50} which means the particle size where the probability of falling into the coarse or fine region is the same. In several cases it seems to be reasonable to take the point of inflexion as a distinguished fictitious particle size d_* (e.g. if $a>0.5$).

The definition of the characteristic T(d) of separation, i.e. of the *performance (efficiency) curve*, is given by

$$T(d) = \frac{\text{weight fraction of size } d \text{ in coarse stream}}{\text{weight fraction of size } d \text{ in feed stream}}$$

It is seen in Fig. 6-1 that the real characteristics usually have size-independent parts indicated by a and b.
By a simple normalization based on

$$T(d) = (1-a-b)\, q(d) + a \qquad (6.1)$$

we get the size-dependent characteristic q(d) which fulfils the boundary conditions $q(0)=0$ and $q(\infty)=1$. Accordingly, the complete characteristic of a classifier consists of size-independent parameters a, b and a size-dependent part q(d), where

$$q(d) = \frac{\text{size-dependent weight fraction of size } d \text{ in coarse stream}}{\text{size-dependent weight fraction of size } d \text{ in feed stream}} . \qquad (6.2)$$

The value of b is, as a rule, negligibly small.

For the approximation of the efficiency curves several functions have been proposed.

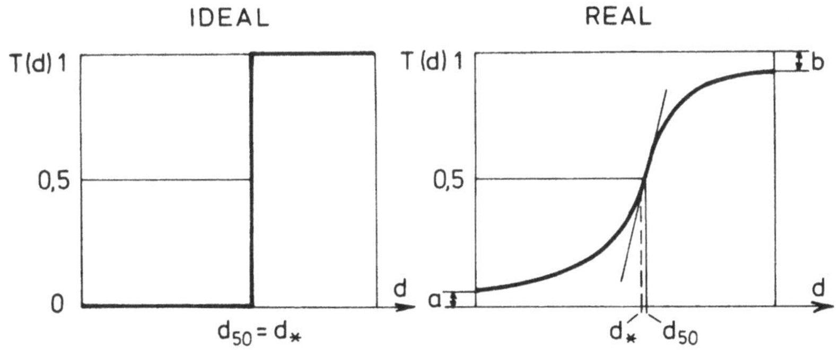

Fig. 6-1 Typical **TROMP** curves of ideal and real classifiers

DRAPER and **LYNCH** in [54] suggested a three-segment polygonal curve on a logarithmic scale.

VAILLANT in [223] proposed the following (non-continuous) function

$$T(d) = \begin{cases} 1-(1-a)\, e^{-\beta\left(\frac{d}{d_0}-1\right)} & d > d_0 \\ a \qquad (\beta > 0) & d < d_0 \end{cases} \tag{6.3}$$

(Now $b \equiv 0$).

AUSTIN and **LUCKIE** in [16] used the curve

$$T(d) = \begin{cases} 1-(1-a)\, e^{-\left(\frac{d_0}{d}\right)^{\beta}} & d > d_0/2 \\ a \qquad (\beta > 0) & d < d_0/2 \end{cases} \tag{6.4}$$

($b \equiv 0$).

For the normalized size-dependent characteristic **LYNCH** in [141] suggested the function

$$q(d) = \frac{e^{\beta\left(\frac{d}{d_0}\right)} - 1}{e^{\beta\left(\frac{d}{d_0}\right)} + e^{\beta} - 2} \qquad (\beta>0) \tag{6.5}$$

where $d_0 = d_{50}$.

Based on the theoretical calculations of **MOLERUS** in [145], **HOFFMANN** proposed the following expression in [172]:

$$q(d) = \frac{1}{1 + \left(\frac{d_0}{d}\right)^2 e^{\beta\left[1-\left(\frac{d}{d_0}\right)^2\right]}} \qquad (\beta>0) \ . \qquad (6.6)$$

ASO introduced the characteristic based on a log-normal distribution (see [172]):

$$q(d) = \frac{1}{\sqrt{2\pi}\sigma} \int_{-\infty}^{d} e^{-\frac{(\ln x - \ln d_0)^2}{2\sigma^2}} \, d(\ln x) \ . \qquad (6.7)$$

The proposal of **PLITT** and **REID** in [164] and [176] represents the **RR** distribution itself:

$$q(d) = 1 - e^{-\left(\frac{d}{d_0}\right)^{\beta}} \qquad (\beta>0) \ . \qquad (6.8)$$

Applying the distribution given by **BERGSTRÖM-GAUDIN-MELOY, HARRIS** in [80] suggested

$$q(d) = 1 - \left[1 - \left(\frac{d}{d_0}\right)^{\beta}\right]^{\gamma} \qquad (\beta,\gamma>0) \ . \qquad (6.9)$$

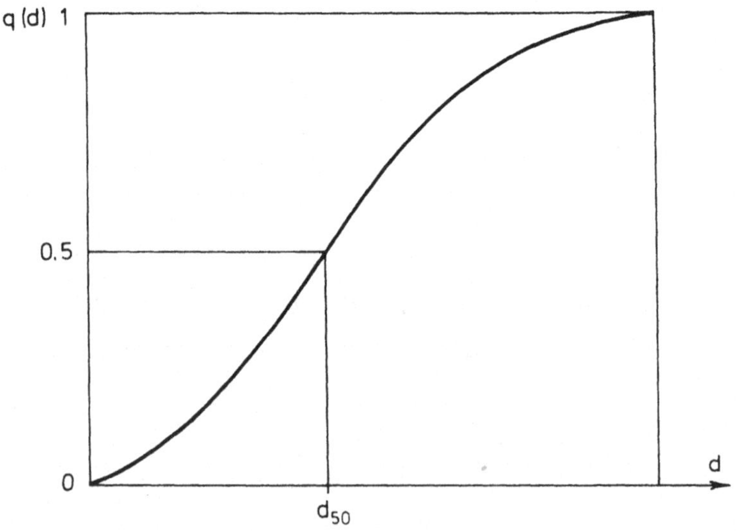

Fig. 6-2 Normalized form of the separation curve of the classifier

Here $d_0 = d_{max}$, i.e. particle of maximum size.

A further suggestion comes from **AUSTIN** in [15]

$$q(d) = \frac{1}{1 + \left(\frac{d_0}{d}\right)^{\beta}} \qquad (\beta > 0) \qquad (6.10)$$

where $d_0 = d_{50}$.

Thus the literature has several, not uniform, suggestions (see [172]); the only common feature is that all the authors try to approximate the normalized efficiency curve by a two-parameter function

$$q(d) = f(\beta, \gamma, d/d_0) \qquad (6.11)$$

and for d_0 usually d_{50} is chosen as reference. Since, in this case, the size-independent part is separated, for the function q(d) the parameter d_{50} always has a meaning (see Fig. 6-2).

As a matter of fact, it is not a difficult task to fit the above nonlinear functions to the measured data by a linear or nonlinear regression method and many applications of such a kind can be found in the literature because of the large number of models.

However, the number of references is rather less, where the dependence of the efficiency curves on the feed, revolutions per minute of the spreading plate in the classifier or any other technological variables - usable for control - is discussed. Essentially, only **AUSTIN** and **LUCKIE** have published results, in [21], based on experient, but they can well be used. In the following, using these results, the control engineering model of the classifier will be derived.

Fig. 6-3 shows mathematically the scheme of the flow of the classifier. The mass balance for the material flow is

$$M = U + G \qquad (6.12)$$

and for the single fractions is

$$M\, D_M'(d) = U\, D_U'(d) + G\, D_G'(d)$$

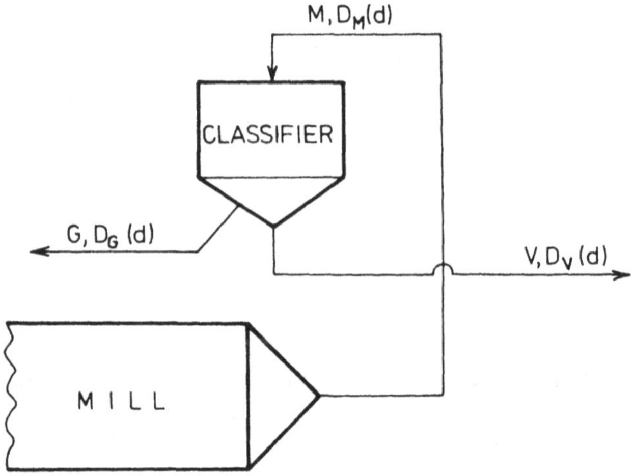

Fig. 6-3 Block-scheme of the material flows of the classifier

as well as

$$M\,D_M(d) = U\,D_U(d) + G\,D_G(d)\ . \tag{6.13}$$

Here $D_M(d)$, $D_V(d)$ and $D_G(d)$ are the (passing) size distributions of the feed of the classifier (mill output), of the final (fine) product and of the grit (coarse), respectively, while $D_M'(d)$, $D_V'(d)$ and $D_G'(d)$ represent the derivative curves corresponding to the density functions. From the last equation we get the expression of **KOULEN-SCHNEIDER** given in [130]

$$D_M(d) = \frac{U\,D_U(d) + G\,D_G(d)}{M} = \frac{D_U(d)+(K-1)D_G(d)}{K} \tag{6.14}$$

where K=M/V is the recycling rate.

According to the definition, the classification characteristic corresponding to the **TROMP** curve can be obtained as

$$T(d) = \frac{GD_G'(d)}{MD_M'(d)} = \frac{M-U}{M}\,\frac{D_G'(d)}{D_M'(d)} = \frac{K-1}{K}\,\frac{D_G'(d)}{D_M'(d)}\ . \tag{6.15}$$

Here we do not intend to deal with the operation of wedge-wire screens.

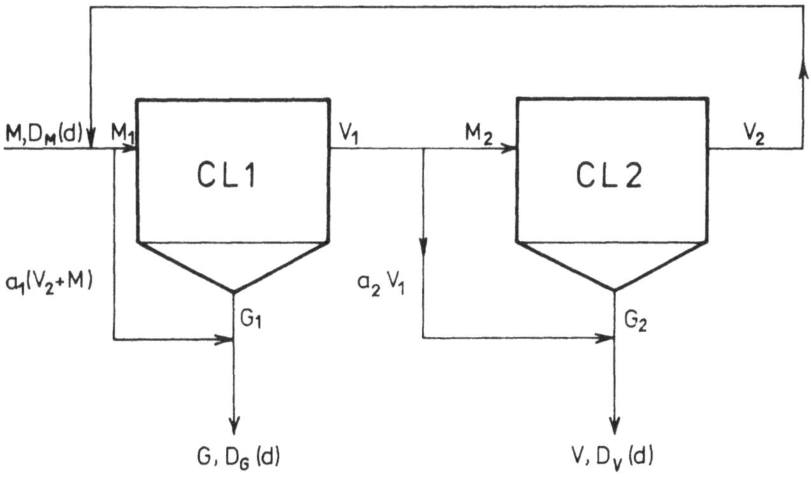

Fig. 6-4 Inner material flows of the classifier

AUSTIN and **LUCKIE** in their model could describe the classification characteristic with fairly good accuracy by applying two serially connected classifiers in cooperation. These inner two classifiers, CL1 and CL2 are needed because volume V_2 of the fine particles is recycling together with the air, but a certain part of the feed (a_1-times) is fed through practically without any classification. In Fig. 6-4 the relation between inner material flows broken down according to the above principle is shown. The relevant mass balance equations are

$$G = G_1 + (M+U_2)\, a_1 \tag{6.16}$$

and

$$U = G_2 + U_1\, a_2 \, . \tag{6.17}$$

The recycling rates for the whole system and for the subsystems are

$$K = \frac{M}{U} \tag{6.18}$$

$$K_1 = \frac{M+U_2}{U_1} \tag{6.19}$$

$$K_2 = \frac{U_1}{U} \, . \tag{6.20}$$

Let us first determine the **TROMP** curves for the two inner fictitious classifiers. By the definition

$$T_2(d) = \frac{U D_U'(d)}{U_1 D_{U1}'(d)} = \frac{1}{K_2} \frac{D_U'(d)}{D_{U1}'(d)} \tag{6.21}$$

as well as

$$T_1(d) = \frac{G D_G'(d)}{M D_M'(d)+U_2 D_{U2}'(d)} = \frac{(K-1)\ D_G'(d)}{D_M'(d)+(K_2-1)D_{U2}'(d)} \ . \tag{6.22}$$

The probabilities of falling into the fine fractions are obtained as

$$1-T_1(d) = \frac{U_1 D_{U1}'(d)}{M D_M'(d)+U_2 D_{U2}'(d)} = \frac{K_2 D_{U1}'(d)}{K D_M'(d)+(K_2-1)D_{U2}'(d)} \tag{6.23}$$

and

$$1-T_2(d) = \frac{U_2 D_{U2}'(d)}{U_1 D_{U1}'(d)} = \frac{K_2-1}{K_2} \frac{D_{U2}'(d)}{D_{U1}'(d)} \ . \tag{6.24}$$

Let us write the complete classification characteristics of the classifier as

$$T(d) = \frac{G D_G'(d)}{M D_M'(d)} = \frac{G D_G'(d)}{M D_M'(d)+0} = \frac{G D_G'(d)}{M D_M'(d)+U_2 D_{U2}'(d)-U_2 D_{U2}'(d)} =$$

$$= \frac{G D_G'(d)}{M D_M'(d)+U_2 D_{U2}'(d)-U_1 D_{U1}'(d)\ \dfrac{U_2 D_{U2}'(d)}{U_1 D_{U1}'(d)}} =$$

$$= \frac{\dfrac{G D_G'(d)}{M D_M'(d)+U_2 D_{U2}'(d)}}{1 - \dfrac{U_1 D_{U1}'(d)}{M D_M'(d)+U_2 D_{U2}'(d)}\ \dfrac{U_2 D_{U2}'(d)}{U_1 D_{U1}'(d)}} = \frac{T_1(d)}{1-[1-T_1(d)][1-T_2(d)]} \ . \tag{6.25}$$

In Fig. 6-4 a_1 and a_2 show the size-independent parts of the classification curves of the two classifiers introduced in Fig. 6-1.

The curves $T_1(d)$ and $T_2(d)$ can be obtained (assuming b=0) as

$$T_1(d) = a_1 + (1-a_1)\, q_1(d) \tag{6.26}$$

and

$$T_2(d) = a_2 + (1-a_2)\, q_2(d) \; . \tag{6.27}$$

AUSTIN and **LUCKIE** were able to determine by measurements the dependence of the normalized separation characteristics $q_1(d)$ and $q_2(d)$, as well as a_1 and a_2 on the feed and the r.p.m. of the spreading plate. They found the best the form of **TROMP** curves corresponding to log-normal distribution frequently used in the literature. In this way they could eliminate the deficiency that the simple log-normal distribution is not able to describe the behaviour represented by the left-hand side of the curve (bending upwards). The phenomenon due to a certain incompleteness of the classifiers could be modelled by the fictitious classifier CL2 introduced in Fig. 6-4.

The log-normal distribution according to (6.7) can be also written as

$$q(d) = \frac{1}{6\sqrt{2\pi}} \int_{-\infty}^{d} \exp \{- \frac{(\ln x - \ln d_{50})^2}{26^2} \} \, d(\ln x) \tag{6.28}$$

and, using an appropriate transformation

$$u = u(x) \; , \tag{6.29}$$

we get

$$q(d) = \frac{1}{\sqrt{\pi}} \int_{-\infty}^{y} e^{-u^2} \, du \tag{6.30}$$

by the substitution

$$u = \frac{\ln x - \ln d_{50}}{6\sqrt{2}} \; . \tag{6.31}$$

(See the meaning of d_{50} in Fig. 6-1.)

The integral in (6.30) cannot be solved in an explicit form; its numerical calculation is possible by expanding in series:

$$q(d) = \frac{1}{\sqrt{\pi}} \left(y - \frac{y^3}{3} + \frac{1}{2!}\frac{y^5}{5} - \frac{1}{3!}\frac{y^7}{7} \pm \dots \frac{y^{2n+1}}{n!(2n+1)} \pm \dots \right) \tag{6.32}$$

(The above Gaussian-integral of errors used to be given in the form of a table but for simulation purposes (6.32) seems to be more suitable and the estimation of error is also simpler, being an alternating series.) For a discrete size distribution having particles in geometrical progression, the upper limit of the integral can be written as

$$y = y_i = \frac{\ln d_i - \ln d_{50}}{6\sqrt{2}} = \frac{1}{6\sqrt{2}} \left(\frac{\ln d_i - \ln d_o}{\ln a} - \frac{\ln d_{50} - \ln d_o}{\ln a} \right) =$$

$$= \frac{1}{6\sqrt{2}} (i - i_{50}) = \beta(i - i_{50}) \,, \tag{6.33}$$

where i is the number of a given screen (or fraction index) and i_{50} is that of belonging to the particle size parameter d_{50}.
Namely,

$$d_i = a^i d_o \tag{6.34}$$

i.e.,

$$d_{i_{50}} = a^{i_{50}} d_o \ . \tag{6.35}$$

The quantity

$$\beta = \frac{1}{6\sqrt{2}} \tag{6.36}$$

introduced here was used by **AUSTIN** to characterize the slope of the **TROMP** curve which is inversely proportional to the variance of a normal distribution.

It follows simply that

$$\left.\frac{\partial q(d)}{\partial u}\right|_{u=0} = \left.\frac{\partial q(d)}{\partial u}\right|_{d=d_{50}} = \frac{1}{6\sqrt{2\pi}} = \frac{\beta}{\sqrt{\pi}} \tag{6.37}$$

According to the above, the normalized classification characteristic for the i-th fraction becomes

$$q_i = q(d_i) = \frac{1}{\sqrt{\pi}} \int_{-\infty}^{\beta(i-i_{50})} e^{-u^2} du = \frac{1}{\sqrt{2\pi}} \int_{-\infty}^{\beta(i-i_{50})} e^{-u^2/2} du . \tag{6.38}$$

AUSTIN and **LUCKIE** in their investigations used the value $a=1/\sqrt[4]{2}$ for discretization and gave the characteristics of the classifiers by β_1, a_1, $i_{50,1}$ and β_2, a_2, $i_{50,2}$,respectively. According to their tests, the parameters of the classifier CL2 were practically independent of the feed M (mill output) and of the r.p.m. of the spreading plate C, i.e. they were taken as

$$\beta_2 \approx 0.855$$

$$a_2 \approx 0.7$$

$$i_{50,2} \approx 18 .$$

This last value corresponds to $d_{50,2}$=9 μm. (The tests were performed by grinding clinker.)

For classifier CL1

$$\beta_1 \approx -0.48$$

was obtained, while the size-independent part of the characteristic T_1 depended on the control variables according to

$$a_1(M,C) = 0.04 + 0.0105 \sqrt{C-250} \log M . \tag{6.39}$$

The dependence of the fraction index $i_{50,1}$ can be written as

$$i_{50,1} = \gamma_1-(\gamma_1-\gamma_2)(1.580-0.5946M+0.0518M^2) , \tag{6.40}$$

where

$$\gamma_1 = \frac{C-200}{23+0.0676(C-200)} \tag{6.41}$$

and

$$\gamma_2 = \frac{C-200}{23+0.0676(C-350)} - 1 \quad . \qquad (6.42)$$

In the evaluation of the results one should remember that the tests were performed on a small cyclone in the laboratory and that generalization requires caution and care. The greatest load of the mill was 7 t/h; the corresponding value in the Cement Works of Hejöcsaba (Hungary) is M=450 t/h. The r.p.m. of the slotting plate was in compliance with that usual in industrial practice (1000 to 1400 r.p.m.).

Using these values, Eq. (6.39) becomes

$$a_1(M,C) = -1.6783 + 0.0105\sqrt{C-250}\ \log M \ , \qquad (6.43)$$

Eq. (6.40) will be

$$i_{50,1} = \gamma_1-(\gamma_1-\gamma_2)(1.580-0.011375M+1.8959\times10^{-5}M^2) \ . \ (6.44)$$

Let us investigate whether the model equations obtained cover current practical knowledge regarding the classifiers. On the basis of (6.44) the maximum of $i_{50,1}$, i.e. the minimum of $d_{50,1}$ is obtained at a mill load of

$$M_0 = \frac{0.011375}{2\ 1.8959\times10^{-5}} \approx 300\ [t/h] \ .$$

This means that on increasing the feed of the classifier - assuming a constant size distribution -, the larger part of this increment gets into the grit for a certain period; after having reached a certain limit, it gets to the final product. This phenomenon is in full agreement with practice, but it should be mentioned that the classifiers are, as a rule, oversized so only the first phenomenon can be observed.

According to (6.43), by increasing the feed of the classifier the asymptotic value of a_1 also increases, so that the sharpness of classification becomes worse.

The influence of the r.p.m. of the spreading plate and the feed M on the classification is summarized in Fig. 6-5.

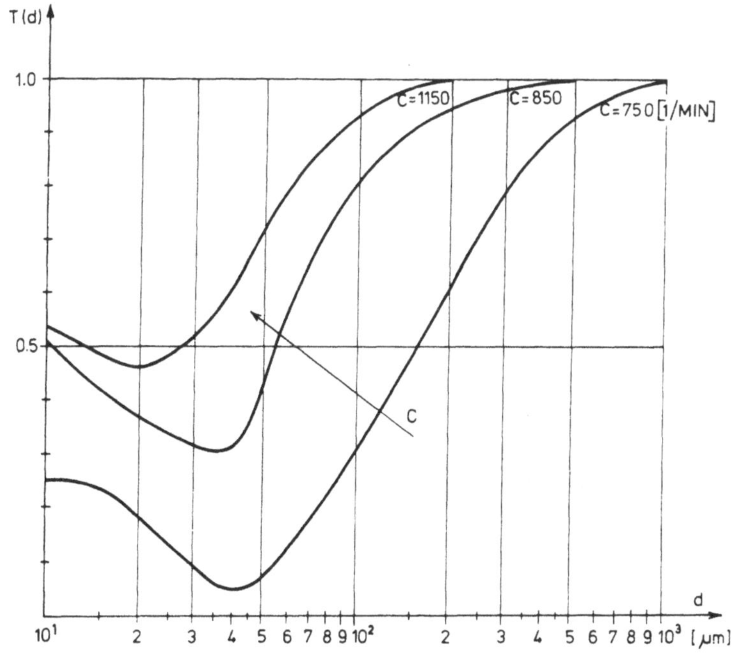

Fig. 6-5/a Effect of the r.p.m. C of the spreading-plate on the **TROMP** curve

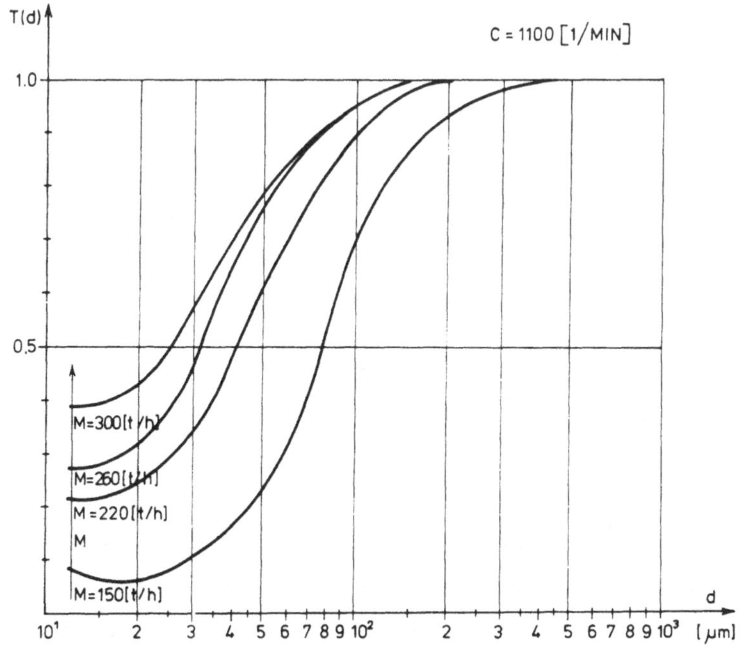

Fig. 6-5/b Effect of feed M on the **TROMP** curve

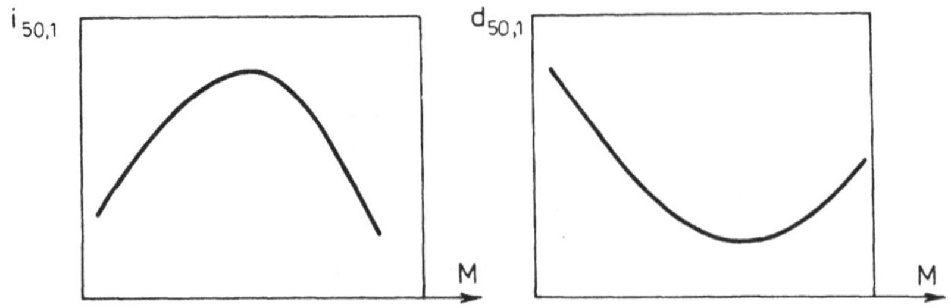

Fig. 6-5/c Change of d_{50} with respect to the feed M

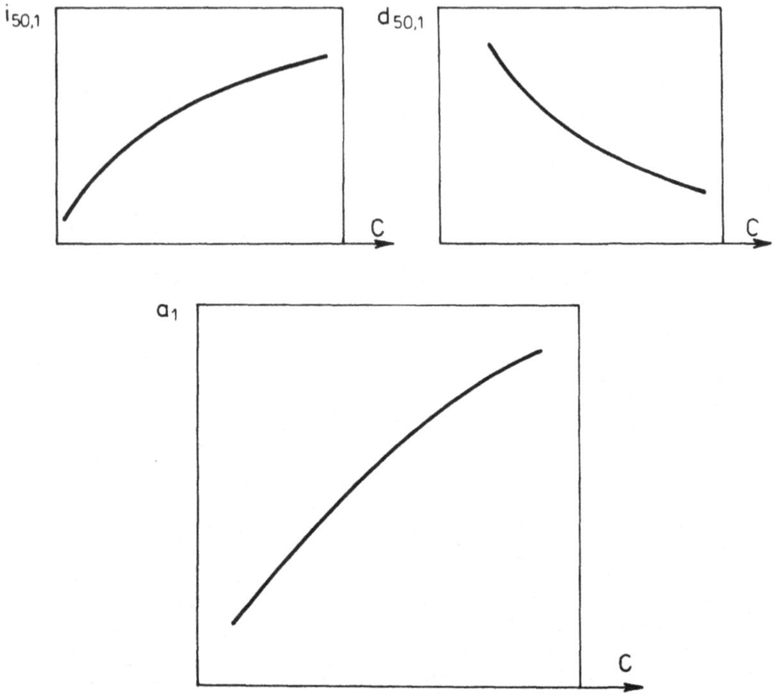

Fig. 6-6 Effect of the r.p.m. C of the spreading-plate on the parameters of the **TROMP** curve

Using (6.39) and (6.40), the effect of the r.p.m., C, for the parameters of the **TROMP** curve is shown in Fig. 6-6. Thus, increasing the r.p.m. causes d_{50} to decrease, but the volume of fine fraction getting to the size-independent grit increases. This means that M and C have practically the same effect, but $d_{50,1}$ has a minimum with respect to M. (Note that for very high r.p.m. - occurring very rarely in practice - a_1 is not a monotonous function of C any more; it has maximum.)

This model of the classifier is well suited to describe the matrix **T** of the dynamic size distribution model (Fig. 5.2-2). In the general quantitative mill model (Fig. 5.1-5) of the macrostructural approach, the effect of r.p.m. can be included in the nonlinearity N_2. On the basis of the experiments in connection with the previous model, the effect of C in (5.1-20) can be taken into account as

$$V(M, \frac{R}{X}, C) = V^o(M)[1-\beta_V(\frac{R}{X} - \frac{R_o}{X_o})-\beta_C(C-C_o)] \quad . \qquad (6.45)$$

For the case of ideal separation and RR distribution, **BEKE** characterized the closed-circuit grinding with the following expressions. Accepting the validity of the **RITTINGER** and **RR** formulae, the relationship between the mill product (M), the aperture of the sieve (d) and the retaining material (R(d)) can be derived simply for open-circuit, continuous, grinding:

$$R(d) = e^{-(\frac{kd}{M})^m} \quad . \qquad (6.46)$$

Here k is an appropriate constant. If, for closed-circuit grinding, an ideal separation is assumed at particle size h, then the relationship between the final product and the mill product is obtained in the form of

$$V = M\,[1 - e^{-(\frac{kh}{M})^m}] = V(M) \quad . \qquad (6.47)$$

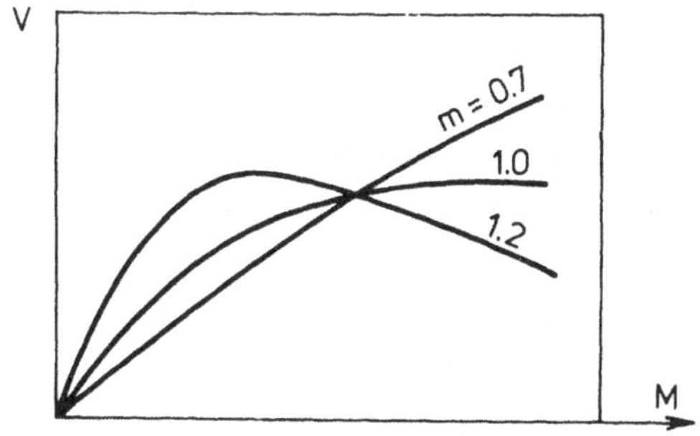

Fig. 6-7 Effect of the smoothing factor on the function V(M)

(Obviously now $h=d_{50}$.) The shape of the curve V(M) is determined by the module m according to Fig. 6-7. It has to be mentioned as a criticism of this relationship that, after classification, the mixture of the ground materials does not follow the **RR** distribution any more, so the above equations are fairly rough approximations of the real situation. According to Fig. 6-7 a small value for m and a large value for the mill load are desired [33]. We may agree with the requirement of higher load for the classifier, since the same is obtained from the previous investigations, namely the sharp classification characteristic requires an oversized classifier. However, the statement regarding the module m is not correct. Besides the fact that for mixed (i.e. classified) ground materials the **RR** distribution is not valid, in practice the assumption $m \geq 1$ always has to be made, otherwise the maximum of the density functions D'(d) will appear at zero, which is nonsense. It is quite a different matter that within a range of possible particle sizes perhaps $m<1$ gives the best approximation locally. However, in practice a most probable particle size different from zero must always exist, which involves $m>1$. Thus, even in case of ideal classification, the curves in Fig. 6-7 must fold back, but it is possible that this phenomenon occurs at an extremely high load only, so it cannot be observed in the majority of cases. Take into consideration that this image is already in better conformity with the model of **AUSTIN** and **LUCKIE.**

As can be seen, we have managed to involve all the essential properties of closed-circuit grinding in our model, including the classification characteristic. In computer simulation models based on the above equations the particle structure of the material flows is represented by vectors derived from discrete distributions. Although, this conception yields a very good description of reality, we should not disregard the fact that technological practice relies much more upon the distribution tests made from the retaining functions R(d) than the complete size distribution investigations performed in a far more complicated technique, of which sedimentation is the final process. The retaining functions can be obtained from the vector density functions of the model in the following way:

$$R(d_i) = 1 - \sum_{j=i}^{n} D'(d_j) = 1 - D(d_i) \quad . \qquad (6.48)$$

As will be seen later, the dependence of the retaining functions on the grinding process can be also obtained quite well from experiments, but it must not be forgotten that the retaining material does not fully represent the distribution. The remaining parameters of the distribution can be defined by fixing the type of distribution. Fig. 6-8 shows, for the **RR** distribution, how the fractions of 3-30 μm and 0-3 μm, which have importance in cement grinding, depend on the retaining function R(90), belonging to the sieve of aperture size 90 μm.

Based on (6.12) and (6.13) the mass balance equations for the retaining functions can be also written for the mill input as

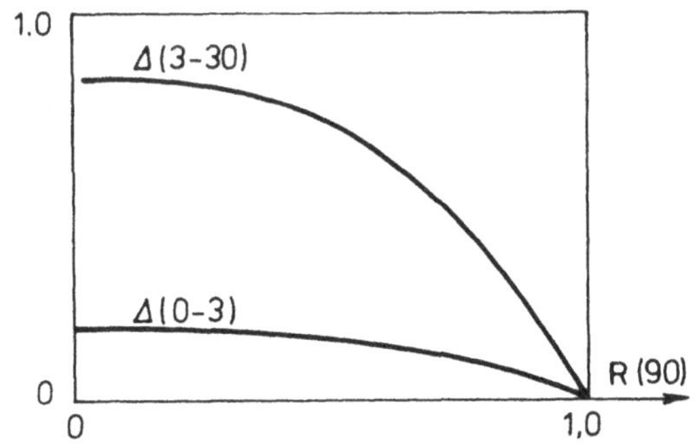

Fig. 6-8 Dependence of the size fractions on the retaining value R(90) (for **RR** distribution) influencing significantly the quality of cement

$$M\ R_X(d) = X\ R_X(d) = R\ R_R(d) + G\ R_G(d) \qquad (6.49)$$

and for the classifier as

$$M\ R_M(d) = U\ R_U(d) + G\ R_G(d) \quad . \qquad (6.50)$$

SCHRAMM determined the characteristics

$$R_M(90) = f_v(M, R_X(90))$$

$$R_U(90) = f_v(M, R_M(90))$$

and

$$R = f(M, R_M(90)) = U$$

by empirical methods using simple polynomial regression. The equations obtained are

$$R_M(90) = -53.42 + 6.64\ M + 0.67\ R_X(90) \qquad (6.51)$$

$$R_U(90) = 12.51 - 7.51\ M + 0.14\ R_M(90) + 1.22\ M^2 \qquad (6.52)$$

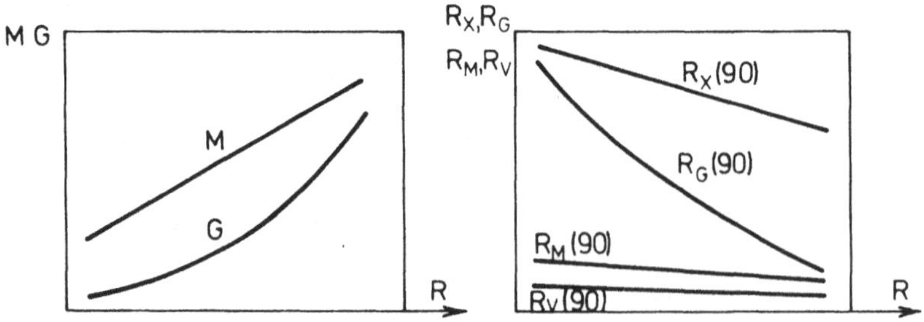

Fig. 6-9 Change of the mill characteristics with respect to the fresh feed given by **SCHRAMM**

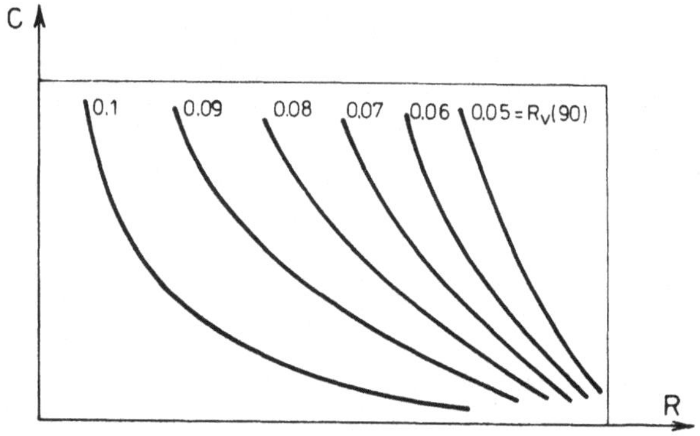

Fig. 6-10 Retaining values R(90) belonging to different feed and r.p.m. of the spreading-plate (according to **SCHRAMM**)

and

$$R = U = 0.84 + 0.15\,M - 0.022\,R_M(90) + 0.051\,M^2 \,. \qquad (6.53)$$

Fig. 6-9 shows the curves obtained as a function of fresh feed R. (Naturally, other relationships can also be given.) It can be stated that the obtained curve sections are quite far from the theoretically possible optimal working point. (Unfortunately these results have also been obtained in the laboratory, so their generalization requires special care.)

SCHRAMM also included the r.p.m. of the spreading plate in his investigations, and his empirical results are summarized by a set of curves in Fig. 6-10. Note that, according to the Figure, the same value of the retaining function can be obtained at different values of R and C [196].

Problems of modelling the specific surface of the final product

In the consideration of the specific surface S in the macrostructural model, i.e. similarly to (6.45), the approximation of the relationship with certain nonlinear function relations depending on M, R/X and C, seems to be a very difficult problem, although several trials have been made in this respect [124],[125].

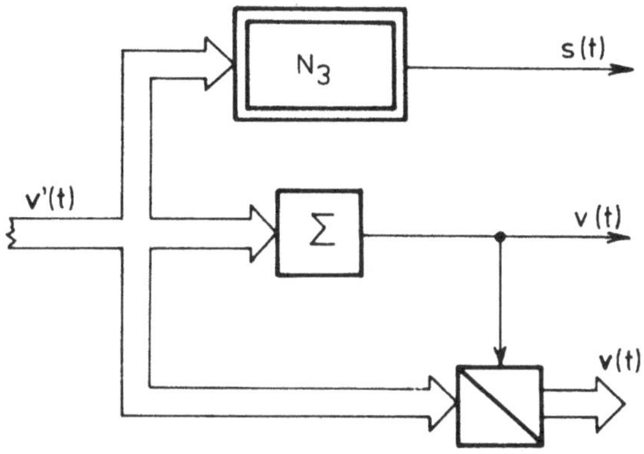

Fig. 6-11 Part of a mill model suitable for calculation of the specific surface of the final product

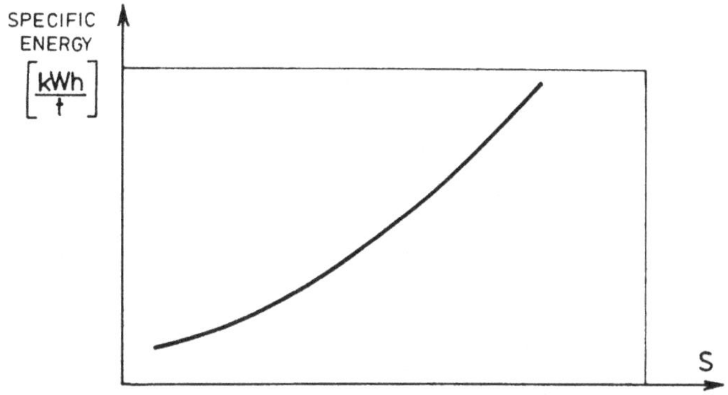

Fig. 6-12 Relationship between the specific surface and the necessary specific energy

The situation is slightly easier with the dynamic size distribution models using discrete density vectors. The model of Fig. 5.2-2 has to be modified with the section of the model shown in Fig. 6-11 in order to include the nonlinear characteristic N_3 as

$$S = \frac{6}{\gamma} \sum_{i=1}^{n} v_i \, a^{-i} = \frac{6}{\gamma \, v(t)} \sum_{i=1}^{n} v_i' \, a^{-i} \tag{6.54}$$

formed analogously with (3.56).

We must not forget that the most widely used method for direct measurement, i.e. the **BLAINE** method, does not provide the values that could be calculated theoretically, because the **BLAINE** device cannot detect the finest fractions [37]. It should be noted that, according to Fig. 6-12, the larger the specific surface, the higher is the specific energy consumption - which is quite obvious, since the grinding produces, for the most part, a new surface. At the same time this means that in stationary operation a finer product (larger specific surface) can be produced by higher specific energy consumption, i.e. as a consequence, by a larger quantity of recycling material. With constant R=V, the specific surface can be increased only through C. Otherwise, in a stationary mode, with a constant recycling rate, a final product of larger specific surface can be produced only in lesser quantity.

Summary of Chapter 6

The modelling of classifiers is essentially based on the literature. After having discussed the different functions suggested for the approximation of the classification curve, a survey of different forms of the efficiency curve as a function of control variables is given.

Then the model suggested by **AUSTIN** and **LUCKIE** is dealt with in detail. In this way all the significant properties of closed-circuit grinding are shown to be included in our model. The computer simulation model constructed from the equations takes the particle structure of the material flows into account by vectors derived from discrete distributions.

The applicability of distribution tests based on retaining functions, widely used in grinding technology, and the consideration of the specific surface in macrostructural models are also treated.

Symbol nomenclature of Chapter 6

d	size (diameter) of the particle [μm]
T(d)	performance (efficiency) curve of the classifier (**TROMP** curve)

a,b	size-independent sections of the performance curve
d_{50}	particle size belonging to 50% passing of the **TROMP** curve [μm]
d*	inflexion point of the **TROMP** curve [μm]
q(d)	size-dependent section of the **TROMP** curve
d_0	distinguished particle size [μm]
ß,γ	constants
d_{max}	maximum particle size of the distribution [μm]
D_M,D_V,D_G	passing particle size distributions (indices refer to particular material flows)
D_M',D_V',D_G'	derivative curves of the density functions (indices refer to particular material flows)
K	recycling rate
G_i, V_i, etc.	material flows of the fictitious inner classifiers [t/h]
$T_1(d)$, $T_2(d)$	**TROMP** curves of the fictitious inner classifiers
a_1, a_2	size-independent sections of the **TROMP** curves of fictitious inner classifiers
K_1, K_2	recycling rates relating to the subsystems
$q_1(d)$, $q_2(d)$	normalized efficiency curves of the fictitious inner classifiers
μ	auxiliary variable used with the Gaussian integral of errors
σ	variance of the normal distribution
ß	the slope of the **TROMP** curve according to **AUSTIN**
i_{50}	serial number (fraction index) of the screen belonging to the 50% passing of the **TROMP** curve
$i_{50,1}$,$i_{50,2}$	the above quantity for the two inner classifiers
$ß_V$, $ß_C$, $ß_M$	parameters
ν_1, ν_2	parameter
h	typical particle size at ideal separation [μm]
m	module smooth (uniformity) factor (at **RR** distribution)
d	aperture of the sieve [μm]
R(d)	retaining function
k	constant
$R_X(d)$, $R_R(d)$	retaining functions (indices refer to particular material flows)
S	specific surface [cm^2/g]
γ	specific weight [g/cm^3]

N_3	nonlinear characteristic
M	mill product [t/h]
X	total feed [t/h]
R	fresh feed [t/h]
G	grit [t/h]
C	revolutions per minute of the spreading plate in the classifier

7. DYNAMIC MODELS OF THE CHEMICAL COMPOSITION OF GROUND MATERIALS

In this chapter dynamic models for describing the changes in the chemical composition of the ground material are dealt with. It is assumed that the mill is fed by raw materials of different chemical composition. The determination of the composition model for the raw-meal of cement production is emphasized, but the results can be generalized for other technologies, too.

The most fundamental intermediate of cement production is the raw-meal produced by the raw mills. In most factories the raw-meal requires homogenization (which may be batch or continuous) before it gets into the cement kiln. The raw-meal, as well as the cement, consists of four principal oxides; these are calcium oxide - (CaO), silica- (SiO_2), alumina - (Al_2O_3) and ferric oxide (Fe_2O_3) which constitute 95% of the total. The remainder consists of magnesium-, potassium- and sodium oxides. Relative parameters are generally used to characterize the raw-meal, containing different combination rates of the oxides, by means of which the chemical composition of the clinker produced by the kiln can be determined [239]. The most frequently used relative parameters are:

1. Calcium modulus (KS_t) :

$$KS_t = \frac{100\ C}{2.8\ S + 1.1\ A + 0.7\ F} \tag{7.1}$$

2. Alumina modulus (AM):

$$AM = \frac{A}{F} \tag{7.2}$$

3. Silica modulus (SM):

$$SM = \frac{S}{A+F} \tag{7.3}$$

Throughout, the following abbreviations are used:

$$C - CaO$$
$$S - SiO_2$$
$$A - Al_2O_3 \qquad (7.4)$$
$$F - Fe_2O_3$$

where the letters refer to mass percentages.

(Regarding the other relative parameters, i.e. lime saturation factor, etc., see the literature [101]. Different parameters are preferred in different countries.)

To determine the dynamic composition model, consider now Fig. 7-1, where a system and process engineering model of a serially connected mill and storage silo is presented. The silo - beside the storage, especially in batch operation - serves to smooth the fluctuations in raw-meal composition by homogenization, required for the proper preparation of the cement kiln. In the Figure r(t) is the mill input resulting from the mix of different raw materials, whose composition vector is **w**(t). As the proportioning is performed by belt scales, the components w_i of the composition vector are called scale rates. The relationship

$$1^T\mathbf{w}(t) = \sum_{i=1}^{3} w_i(t) = 1 \qquad (7.5)$$

is obviously fulfilled, because **w**(t) is a composition vector. (Throughout, - in accordance with Hungarian practice -three raw material components are assumed; of course, the results can be easily generalized for components of any number.) The available raw materials can be characterized by a so-called composition matrix

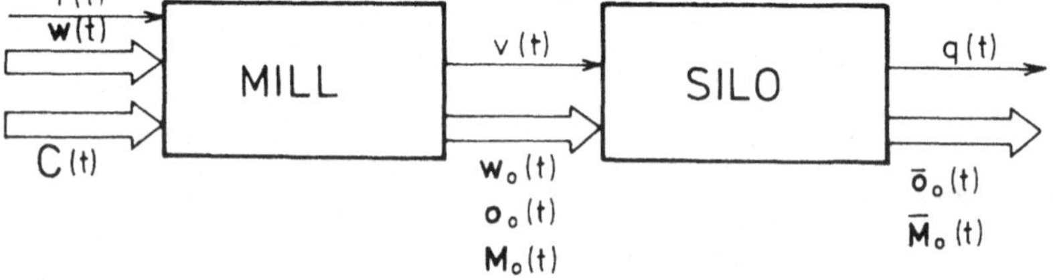

Fig. 7-1 Block-scheme of a mill-silo system

$$C = \begin{pmatrix} C_1 & C_2 & C_3 \\ S_1 & S_2 & S_3 \\ A_1 & A_2 & A_3 \\ F_1 & F_2 & F_3 \end{pmatrix} \tag{7.6}$$

where the columns refer to the corresponding bunkers. The scale rate w_i means that w_i percent. of the mill feed r(t) is taken from the i-th bunker represented by oxide-percentage C_i, S_i, A_i, F_i.

From the i-th component the volume

$$r_i(t) = w_i(t)\ r(t) \tag{7.7}$$

is fed into the mill. The vector of input partial material flows can be given by

$$w_i'(t) = w(t)\ r(t) \quad . \tag{7.8}$$

(The notation ' is in compliance with the concept of the absolute material flow introduced in Chapter 4, i.e. $\mathbf{1}^T w_i'(t) = r(t)$.)
Introduce the vector of oxides characterizing the input material flow r(t) of the mill as

$$o_i(t) = C\ w(t). \tag{7.9}$$

Here and in the following **C** is assumed to be known. (The composition matrix **C** has changing elements, but we calculate using the average of these changes over a longer period. As a consequence, it is more correct to call $\mathbf{C}_0$ the average composition matrix, whose change is, however, much slower than that of the mill transient, so during this modelling procedure it can be taken constant.)

In (7.9) the components of $\mathbf{o}_i(t)$ are the oxide contents in percent., so they give the value C, S, A, F for the complete input material flow.
It seems to be reasonable to introduce the absolute quantities for the oxides, too; i.e., to form fictitious oxide flows. On the analogy of (7.8) the vector of partial oxide flows is given by

$$o_i'(t) = o_i(t)\, r(t) \quad . \tag{7.10}$$

The relationship between the oxide composition and modulus values is determined by the nonlinear equations (7.1) - (7.3). Thus the vector of modulus variables

$$M = [KS_t,\ AM,\ SM]^T \tag{7.11}$$

depends, through a nonlinear vector-vector function relation

$$M = M[o(t)], \tag{7.12}$$

on the vector of the oxide composition

$$o(t) = [C(t)\ S(t)\ A(t)\ F(t)] \quad . \tag{7.13}$$

This relationship can be given for both input and output material flows of the mill and for the output of the silo.

It could be interesting to investigate the inverse of (7.12); i.e., to determine the oxide composition vector ensuring a given modulus value. Introducing the notation μ for the remaining oxides and applying

$$1^T o = 1-\mu \tag{7.14}$$

the following set of equations can be written

$$\begin{pmatrix} -100 & 2.8KS_t & 1.1KS_t & 0.7KS_t \\ 0 & -1 & SM & SM \\ 0 & 0 & -1 & AM \\ 1 & 1 & 1 & 1 \end{pmatrix} \mathbf{o} = \mathbf{G(M)o} = \begin{pmatrix} 0 \\ 0 \\ 0 \\ 1-\mu \end{pmatrix} = \mathbf{g} \tag{7.15}$$

whence

$$o = G^{-1}(M)\, g \quad . \tag{7.16}$$

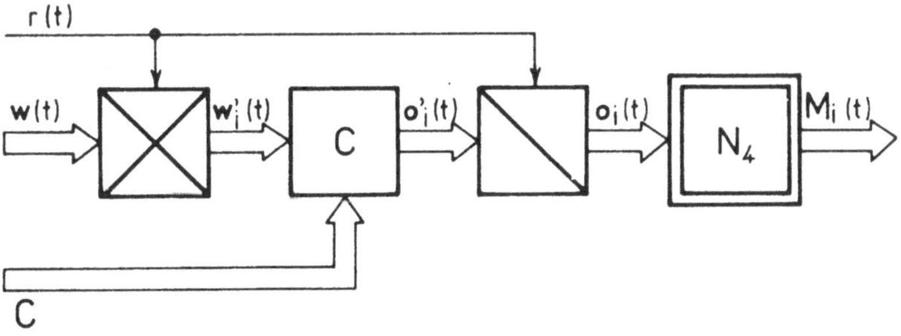

Fig. 7-2 Model of the composition of the input material flow

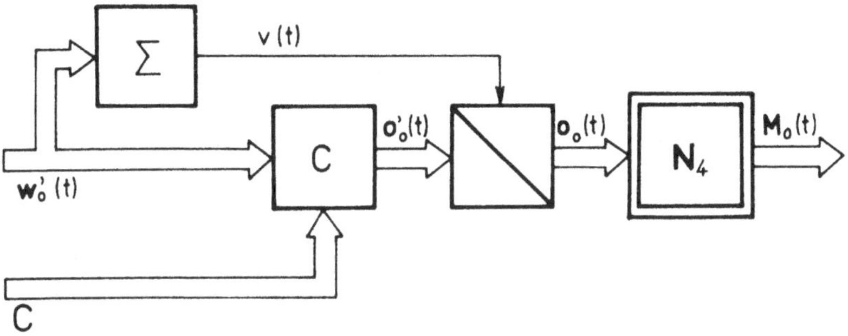

Fig. 7-3 Model of the composition of the output material flow

The structure of matrix **G(M)** is different for any other definition of the modulus!

In the foregoing, the problems of modelling of the composition of input material flow have been summarized and the relationships of the introduced variables are illustrated in Fig. 7-2 for the input material flows. The index i refers to the input here.

Applying the above modelling concepts to the output material flow of the mill, we get Fig. 7-3, where v(t) is the output material flow and for the other variables index o refers simply to the output.

The question now is between which variables the mill establishes a primarily dynamic relation. As is known, the establishment of a grinding model for the mix of materials is very difficult, because the mixture does not have the same properties as those of its components.

A better approach is to assume that the individual input material types go through the mill according to separate dynamic characteristics and the macroscopic phenomena to be observed result from the joint effect of these decoupled characteristics. (It is known that during the start-up transient of the mill, a product enriched in softer components appears at first, then it gradually reaches the composition corresponding to the average hardness and grindability. The inverse of this phenomenon takes place during the stop-down transient of the mill.)

As it turned out from the previous chapters, and in agreement with the dynamic behaviour cited frequently in the literature [38], the transfer function between quantities for the open- and closed-circuit grinding mills can be well approximated by first order lags having dead-time. So we have reason to suppose a similar transfer function between the i-th raw meal $w_{ii}'(t)$ of the input and $w_{oi}'(t)$ of the output material flows as

$$Y_{mi}(s) = \frac{e^{-s\tau_i}}{1+sT_i} . \qquad (7.17)$$

The multi-channel non-coupled linear dynamic model obtained in this way is shown in Fig. 7-4, where the dynamics of the multivariable system are given by the diagonal transfer function matrix

$$Y_m(s) = \mathrm{diag} \left< \frac{e^{-s\tau_1}}{1+sT_1}, \frac{e^{-s\tau_2}}{1+sT_2}, \frac{e^{-s\tau_3}}{1+sT_3} \right> \qquad (7.18)$$

which could be built up of elements of higher order, or of quantitative mill models used in Chapter 5 (three parallel operating models) to achieve higher accuracy.

In practice, this latter solution cannot be realized due to the unavoidable occurrence of interactions and a great number of parameters having no physical meaning.

The multivariable composition model suggested by us is shown in Fig. 7-5. It is obvious from the Figure that the multivariable system is coupled because of the non-diagonal *composition matrix* C.

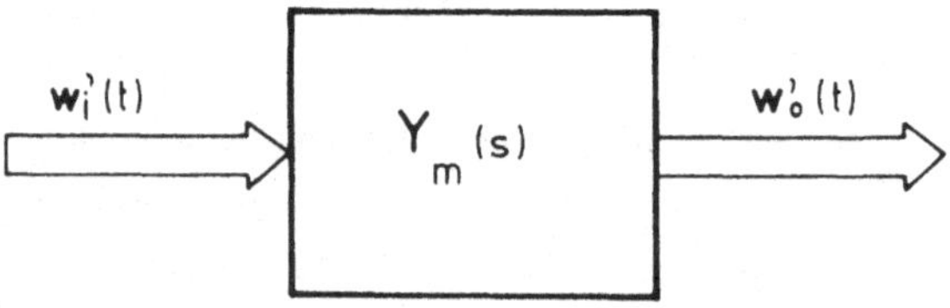

Fig. 7-4 Approximating multivariable linear dynamic composition model of the grinding system

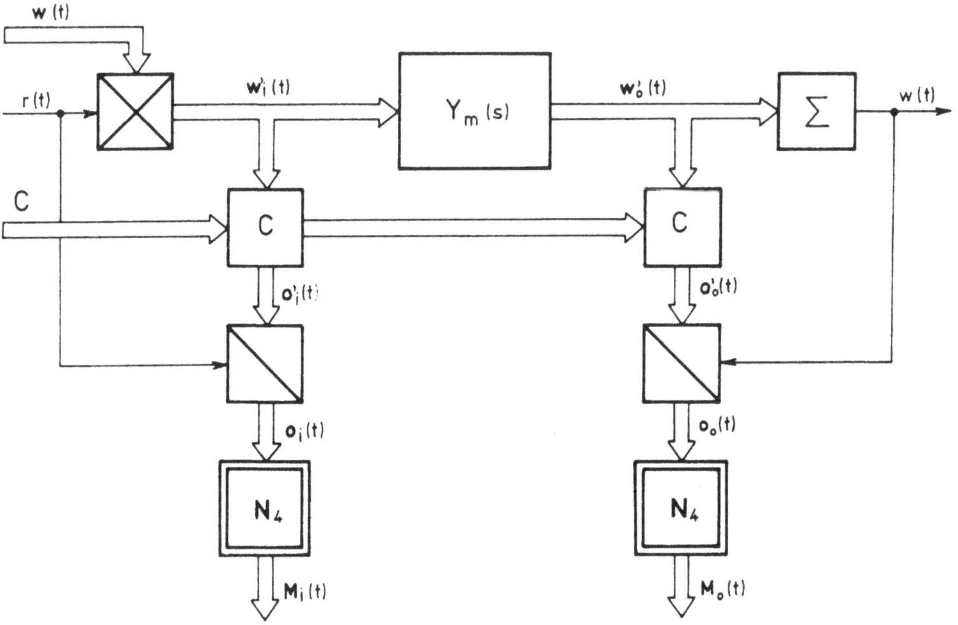

Fig. 7-5 Multivariable coupled composition model of the grinding system

The nonlinearity N_4 of the modulus calculations is taken into consideration by (7.1) - (7.3), in one direction, and by (7.15), in the other. Progressing in this latter direction, the scale rates belonging to a particular modulus can also be determined, and the following set of linear equations is obtained:

$$\begin{pmatrix} C_1 & C_2 & C_3 & -1 & 0 & 0 & 0 \\ S_1 & S_2 & S_3 & 0 & -1 & 0 & 0 \\ A_1 & A_2 & A_3 & 0 & 0 & -1 & 0 \\ F_1 & F_2 & F_3 & 0 & 0 & 0 & -1 \\ 1 & 1 & 1 & 0 & 0 & 0 & 0 \\ 0 & 0 & 0 & -100 & 2.8KS_t & 1.1KS_t & 0.7KS_t \\ 0 & 0 & 0 & 0 & 0 & -1 & AM \end{pmatrix} \begin{pmatrix} w_1 \\ w_2 \\ w_3 \\ C \\ S \\ A \\ F \end{pmatrix} = \mathbf{H(C,M)wo} = \begin{pmatrix} 0 \\ 0 \\ 0 \\ 0 \\ 1 \\ 0 \\ 0 \end{pmatrix} = \mathbf{h} \quad (7.19)$$

where (7.9), (7.5) and two rows of (7.15) are used. (Note that any other two equations of (7.15) could have been used.) Thus the scale rates

$$wo = H(C,M)^{-1}h \quad (7.20)$$

can be determined by matrix inversion. In this case the size of the matrix to be inverted is (7×7). To solve the equation, the composition matrix **C** has to be known.

Write now the first three rows of (7.9) (in general **C** is not quadratic), or any other triplet can be chosen if the equation obtained is a regular one:

$$\begin{bmatrix} C \\ S \\ A \end{bmatrix} = \tilde{o} = \begin{bmatrix} C_1 & C_2 & C_3 \\ S_1 & S_2 & S_3 \\ A_1 & A_2 & A_3 \end{bmatrix} \begin{bmatrix} w_1 \\ w_2 \\ w_3 \end{bmatrix} = \tilde{C}\, w \, . \qquad (7.21)$$

Here $\tilde{o}$ and $\tilde{C}$ are the corresponding truncated vector and matrix, respectively. Thus, on the basis of (7.15) and (7.21) we get

$$w = \tilde{C}^{-1}\tilde{o} = \tilde{C}^{-1}\; G^{-1}(M)g \qquad (7.22)$$

in which, instead of the previous matrix of size 7×7, matrices of size 3×3 nad 4×4 have to be inverted. As has been mentioned earlier, there are several possibilities for solving the equations, but they are not discussed here. It has to be mentioned that, by using a **C** of special structure and under given moduli, very simple solutions can also be obtained with proper care.

It is obvious from (7.19) that in the case of three kinds of raw meals - due to the special construction of the modulus values - only two of the moduli can be adjusted arbitrarily; the third one follows from their value. Taking this into account, the following set of equations with six unknowns seems to be the most correct description while the quantities

$$w_3 = 1 - w_1 - w_2 \qquad (7.23)$$

and

$$SM = \frac{S}{F(AM+1)} \qquad (7.24)$$

are simply obtained.

The system of equations

$$\begin{bmatrix} (C_1-C_2) & (C_2-C_3) & -1 & 0 & 0 & 0 \\ (S_1-S_3) & (S_2-S_3) & 0 & -1 & 0 & 0 \\ (A_1-A_3) & (A_2-A_3) & 0 & 0 & -1 & 0 \\ (F_1-F_3) & (F_2-F_3) & 0 & 0 & 0 & -1 \\ 0 & 0 & -100 & 2.8KS_t & 1.1KS_t & 0.7KS_t \\ 0 & 0 & 0 & 0 & -1 & AM \end{bmatrix} \begin{bmatrix} w_1 \\ w_2 \\ C \\ S \\ A \\ F \end{bmatrix} = \begin{bmatrix} -C_3 \\ -S_3 \\ -A_3 \\ -F_3 \\ 0 \\ 0 \end{bmatrix} =$$

$$= E(C,M)\ w\ \tilde{o} = e \tag{7.25}$$

contains the composition matrix **C**; by substituting the required moduli values into the matrix of coefficients, we can obtain the scale rates to be adjusted and the percentage of oxides to be ensured.

In the model of chemical composition, stochastic disturbances may occur at two places due to the fluctuation of the quality and composition of the raw meals. On the one hand, variations in the grindability may occasionally change the time delays τ_i and time constants T_i, and in practice the composition matrix has the form

$$C = C_o + \Delta C, \tag{7.26}$$

on the other, where C_0 represents the slowly changing average composition and ΔC denotes the stochastic disturbances. From long-term observations actually C_0 can be estimated by different statistical methods, so this value can be assumed for **C** in our equations. Due to the above, control methods which identify the coefficients of the composition matrix or do not assume a priori knowledge of **C** may have special importance.

In the model shown in Fig. 7-5 the fundamental process is described by the relations between partial material flows, so it seems to be obvious to choose r(t) and **w**(t) as the (possibly control purpose) input (intervening) signals. The other variables, including also the output ones, can be regarded as outcomes. Regardless of this model, relationships can be established between the outcomes without taking the real flowchart into consideration, but special care is required.

For example, let $\mathbf{o}_i'(t)$ and $\mathbf{o}_0'(t)$ be the input and output of our fictitious model, respectively. The following relationship can be written between the Laplace transform of the variables:

$$o_o'(s) = Cw_o'(s) = CY_m(s)w_i'(s) = CY_m(s)C^{-1}o_i'(s) \quad (7.27)$$

i.e. this equation has an unambiguous solution only if **C** is quadratic and regular. This procedure, i.e. to develop a simple linear model between $\mathbf{o}_i(t)$ and $\mathbf{o}_o(t)$, cannot be continued, because calculation of these variables requires nonlinear operations. The same is valid for moduli $\mathbf{M}_i(t)$ and $\mathbf{M}_o(t)$.

The above nonlinearities, of course, do not prevent us from developing, for example, a multivariable dynamic model linearized in the working point between $\mathbf{w}(t)$ and $\mathbf{M}_o(t)$.

7.1. Modelling of homogenizing silos

It is seen from the technological sketch in Fig. 7-1 that the raw meal gets into the homogenizing silos. Two kinds of silos, continuous and batch operation, can be distinguished.

For batch silos, the homogenization is performed after the silos are completely filled. The homogenization does not influence the average composition of the silo content, but it decreases - in inverse proportion to the residence time - the fluctuations of the raw meal composition around the average composition determined by the feed. The effectiveness of homogenization decreases with time; moreover, an opposite process may also occur. Therefore, the necessary residence time has to be determined by experiment.
To make the production continuous, storage silos are used.

Continuous silos have much smaller halls and the homogenization effect is ensured by a special method of simultaneous and continuous charge and discharge. The modelling of silos of different types is not discussed here, but the average oxide composition of the silo output is studied.

Batch silo

The instantaneous silo content can be given by the integral

$$q(t) = \int_0^t v(\tau)\, d\tau + q(0) \quad (7.28)$$

where $v(\tau)$ is the material flow entering the silo from the mill. The lower limit of the integral 0 represents the beginning of the filling-up procedure; q(0) is usually zero. The average $\bar{o}_0(t)$ of the oxide composition of the silo content is determined by the equation (for the sake of simplicity, q(0)=0) :

$$\bar{o}_0(t) \int_0^t v(\tau)\, d\tau = \int_0^t v(\tau)\, o_0(\tau)\, d\tau = \int_0^t o_0'(\tau)\, d\tau \qquad (7.29)$$

whence

$$\bar{o}_0(t) = \frac{\int_0^t v(\tau) o_0(\tau)\, d\tau}{\int_0^t v(\tau)\, d\tau} = \frac{\int_0^t v(\tau) o_0(\tau)\, d\tau}{q(t)} \; . \qquad (7.30)$$

The related model of the batch silo is given in Fig. 7-6. In order to create the state space model, (7.29) has to be written in differential form:

$$q(t)\, \frac{d\bar{o}_0(t)}{dt} + \bar{o}_0(t)\, v(t) = v(t)\, o_0(t) \qquad (7.31)$$

whence

$$\frac{d\bar{o}_0(t)}{dt} = \frac{v(t)}{q(t)} \left[\, o_0(t) - \bar{o}_0(t) \right] . \qquad (7.32)$$

The block-scheme of this equivalent model is shown in Fig. 7-7. Another version of (7.32) is

$$\frac{q(t)}{v(t)}\, \frac{d\bar{o}_0(t)}{dt} + \bar{o}_0(t) = o_0(t) \qquad (7.33)$$

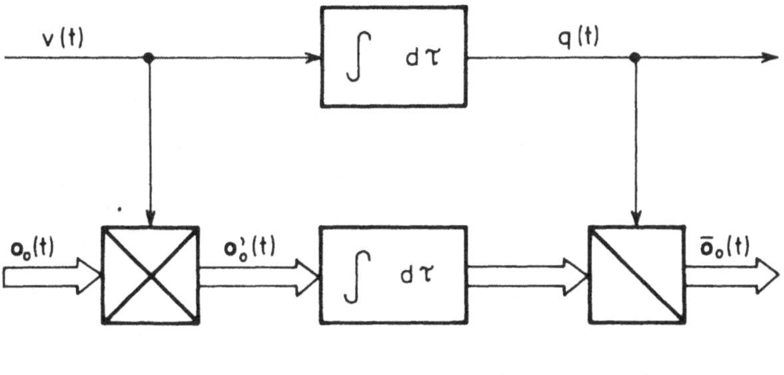

Fig. 7-6 Functional model of a batch homogenizing silo

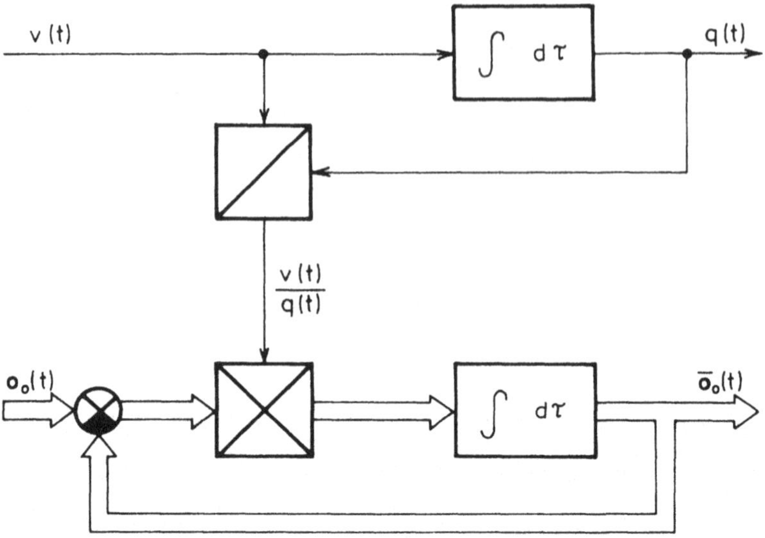

Fig. 7-7 State-space form of the functional model of a batch homogenizing silo

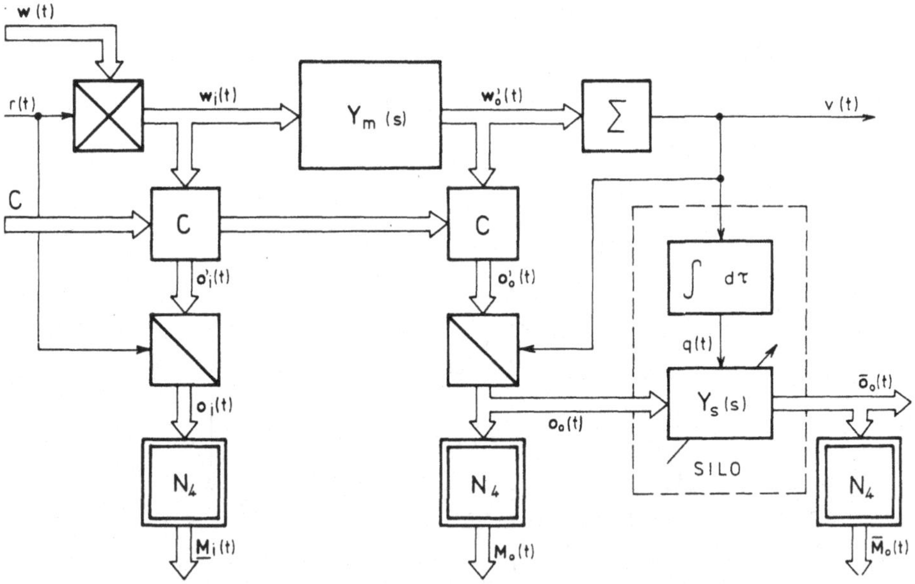

Fig. 7-8 Block-scheme of the composition model of a mill and a batch silo

Fig. 7-9 Block-scheme of a continuous silo

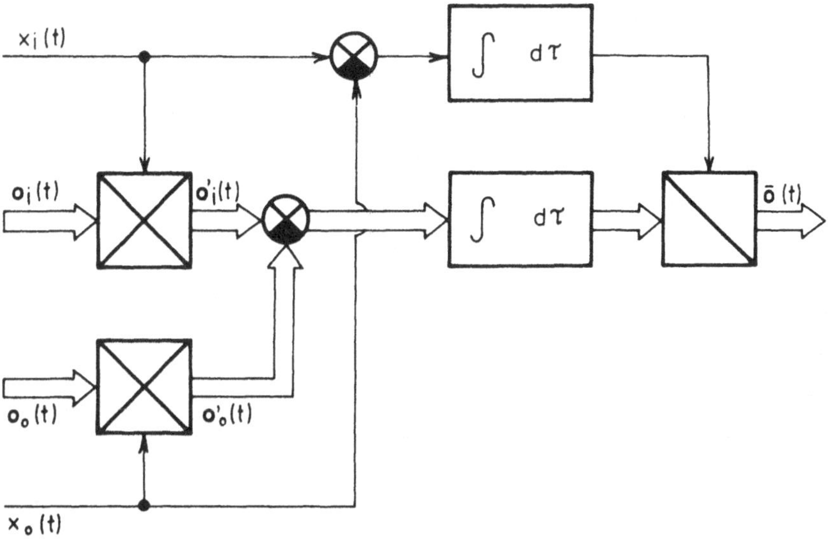

Fig. 7-10 Model of a continuous silo

which is a first-order linear differential equation with time-varying parameters. By formal comparison, we get a first-order lag with time-varying time constant as

$$T = \frac{q(t)}{v(t)} . \tag{7.34}$$

For interest, in Fig. 7-8 the block-cheme of the composition model of a mill and batch silo is illustrated, where $\mathbf{Y}_S(s)$ represents the transfer matrix of the multivariable linear system with varying parameters, defined by (7.33), and $\bar{M}_0(t)$ is the vector of moduli calculated from $\bar{o}_0(t)$.

Continuous silo

For this model - which is far more complicated than that of the batch silo - let us consider Fig. 7-9, where $x_i(t)$, $x_o(t)$ are the input and output material flows, $\mathbf{o}_i(t)$, $\mathbf{o}_o(t)$ are the vectors of the oxide percentages characterizing the chemical composition of the flows. The instantaneous content of the silo

$$q(t) = \int_0^t [x_i(\tau) - x_o(\tau)]\, d\tau + q(0) . \qquad (7.35)$$

The average oxide-composition $\bar{o}(t)$ of the silo content can be given by the equation (for the sake of simplicity, q(0)=0):

$$\bar{o}(t)q(t) = \int_0^t [x_i(\tau)o_i(\tau)-x_o(\tau)o_o(\tau)d\tau] = \int_0^t [o_i'(\tau)-o_o'(\tau)]d\tau \qquad (7.36)$$

whence

$$\bar{o}(t) = \frac{\int_0^t [o_i'(\tau)-o_o'(\tau)]\, d\tau}{q(t)} . \qquad (7.37)$$

The block-scheme of the model relating to this equation is shown in Fig. 7-10. By introducing certain assumptions, this model can be further simplified. In the stationary case

$$X = x_i(t) \equiv x_o(t), \qquad t > T_o \qquad (7.38)$$

so

$$q(t) = q(T_o) = Q = \int_0^{T_o} [x_i(\tau)-x_o(\tau)]\, d\tau . \qquad (7.39)$$

Then

$$\bar{o}(t) = \frac{1}{Q} \int_0^t X\Delta o(\tau)\, d\tau + \bar{o}(T_o) , \qquad (7.40)$$

where

$$\Delta o(t) = o_i(t) - o_o(t) . \qquad (7.41)$$

Let us write (7.36) in differential form:

$$q(t) = \frac{d\bar{o}(t)}{dt} + \bar{o}(t)[x_i'(t)-x_o(t)] = x_i(t)o_i(t)-x_o(t)o_o(t). \qquad (7.42)$$

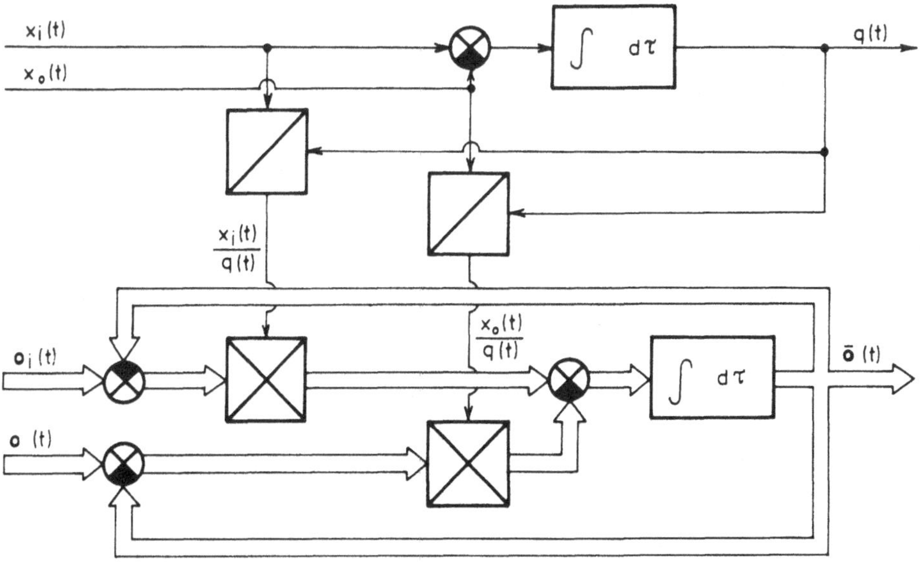

Fig. 7-11 Block-scheme of the simplified state-space model of a continuous silo

This state equation can be also given as

$$\frac{d\bar{o}(t)}{dt} = \frac{x_i(t)}{q(t)}[o_i(t)-\bar{o}(t)] - \frac{x_o(t)}{q(t)}[o_o(t) - \bar{o}(t)], \tag{7.43}$$

the block-scheme of which is shown in Fig. 7-11. For the special case $d\bar{o}(t)/dt=0$, in steady-state the composition of silo content becomes

$$\bar{o} = \frac{x_i(t)o_i(t)-x_o(t)o_o(t)}{x_i(t) - x_o(t)} , \tag{7.44}$$

however, for the condition (7.38), the differential form of (7.40) is obtained as

$$\frac{d\bar{o}(t)}{dt} = \frac{X}{Q}[o_i(t) - o_o(t)] . \tag{7.45}$$

In a very special case, when

$$o_o(t) = \bar{o}(t) \tag{7.46}$$

we get

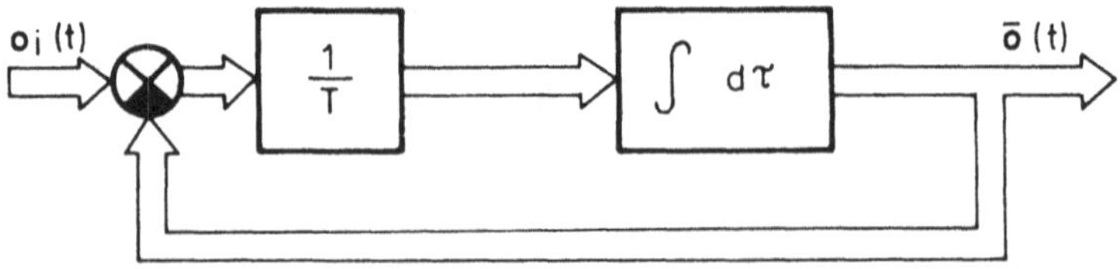

Fig. 7-12 Block-scheme of the first-order differential equation model with constant parameters of a continuous silo

$$\frac{d\bar{o}_0(t)}{dt} = \frac{X}{Q}\,[o_i(t) - o_0(t)] \,. \qquad (7.47)$$

or in another form

$$\frac{Q}{X}\,\frac{do_0(t)}{dt} + o_0(t) = o_i(t) \qquad (7.48)$$

which is very similar to (7.33) but now it is a first-order linear differential equation with constant parameters.

By formal comparison the time constant

$$T = \frac{Q}{X} \qquad (7.49)$$

is obtained, which is actually the duration of silo filling-up. The joint composition model of the mill and the continuous silo is shown in Fig.7-12. Note that the condition (7.46) can be fulfilled only in case of a special silo in- and outlet.

Summary of Chapter 7

In this chapter, dynamic models of the chemical composition of raw meal are considered. The main difficulty is caused by the stochastically changing elements of the composition matrix **C**, and only its long-term average can be taken into account.

After having described the set of equations suitable for modelling the chemical composition of the input material flow, a composition model is derived that includes the dynamic behaviour of the mill relating to partial material flows. Then problems of modelling batch and continuous homogenizations are treated.

Symbol nomenclature of Chapter 7

KS_t	calcium modulus
AM	alumina modulus
SM	silica modulus
C	CaO
S	SiO_2
A	Al_2O_3
F	Fe_2O_3
$\mathbf{w}(t)$	composition vector of the feed [t/h]
$\mathbf{1}^T$	summing vector
$w_i(t)$	input scale rates
$\mathbf{C}$	composition matrix of the raw meal
$r(t)$	input material flow of the mill [t/h]
$\mathbf{w}_i'(t)$	vector of input partial material flows
$\mathbf{o}_i(t)$	vector of oxides relating to the input material flow r(t) of the mill
$\mathbf{C}_o$	average composition matrix of the raw meal
$\mathbf{o}_i'(t)$	vector of partial oxide-flows of the mill input
$\mathbf{o}_o'(t)$	vector of partial oxide-flows of the mill output
$\mathbf{M}$	vector of moduli
μ	remaining oxides
$\mathbf{G}(\mathbf{M})$	coefficient matrix
$v(t)$	output material flow of the mill [t/h]
$w_{ii}'(t)$	input partial material flow related to the i-th raw meal
$w_{oi}'(t)$	output partial material flow related to the i-th raw meal
$Y(s)$	transfer function
s	Laplace operator
τ_i	dead-time [min]
T_i	time constant

N_4	nonlinearity representing the moduli calculations
w	vector of scale rates
H(C,M)	coefficient matrix
h	constant vector
$\tilde{o}$	truncated vector of the oxide variables
$\tilde{C}$	truncated matrix of the composition variables
e	auxiliary vector
ΔC	matrix of stochastic changes in the composition
$v(\tau)$	material flow entering the silo from the mill [t/h]
$q(t)$	instantaneous silo content
$\bar{o}_0(t)$	vector of average oxide composition of the silo content
X	steady-state input/output material flow for continuous homogenization silo [t/h]
Q	degree of filling of the continuous homogenization silo [t]
$\bar{M}_0(t)$	average moduli vector of the silo content.

8. ADVANCED CONTROL SYSTEMS FOR GRINDING

8.1. Composition control

Composition control serves to ensure the desired average chemical composition of the raw meal, i.e. to keep the oxide variables (C, S, A, F), or their relative rates represented by the moduli, to prescribed values. The main problem of composition control derives from the time needed for measurement of the oxide content which causes considerable delay in the closed-loop control system. Its value may be so large (from 30 minutes to one hour) that several experts advocate improvements in the quality of control by decreasing this dead-time rather than by using the most advanced control algorithms. Previously chemical analysis was performed by manual sampling and laboratory procedures. Now, representative sampling is performed automatically and the oxide content is analyzed by an X-ray fluorescence analyzer (RFA). As the time delay caused by the measuring method cannot be significantly reduced further at an acceptable level of cost, so effective control algorithms play a more important role.

The complexity of composition control mainly depends on the number of raw meals to be used. In the majority of cement plants 3 to 5 raw meals are applied, but it may happen that 8 to 10 different raw meals are mixed. The qualitative properties attainable by the control processes depend largely on the variation in oxide content of the raw meals. Occasionally it may even happen that there is no need for closed-loop control at all. However, in many cases the changes are so large that solutions are possible only by proper computer control. This is the reason why computer control first appeared in our technological unit of cement production.

A simplified technological scheme for composition control of the raw meal is shown in Fig. 8.1-1 for the case of the three most frequently used raw meals (limestone, clay and pyrite). The compositions are different in the bunkers. The proportioning is performed by means of computer operated belt scales. For the proper preparation of the cement kiln the raw meal obtained from the (usually closed-circuit) raw mill requires further homogenization in blending silos. A representative sample is taken from the raw meal by a suitable equipment and forwarded through a dispatch-tube to the RFA which provides the oxide composition as the result of the analysis. Usually the entire automatic process is supervised by the computer.

Dynamic models for the chemical composition of the ground material have been discussed in

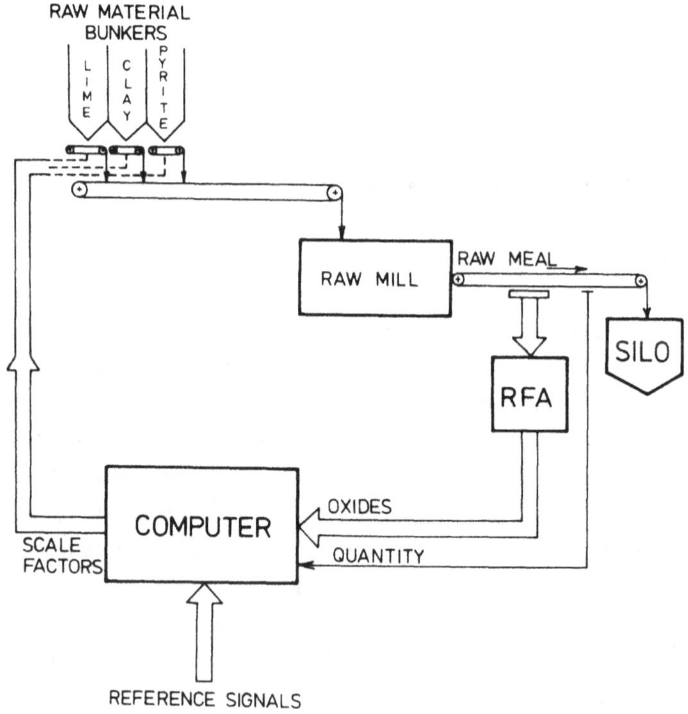

Fig. 8.1-1 Simplified technological scheme of a raw-meal composition control system

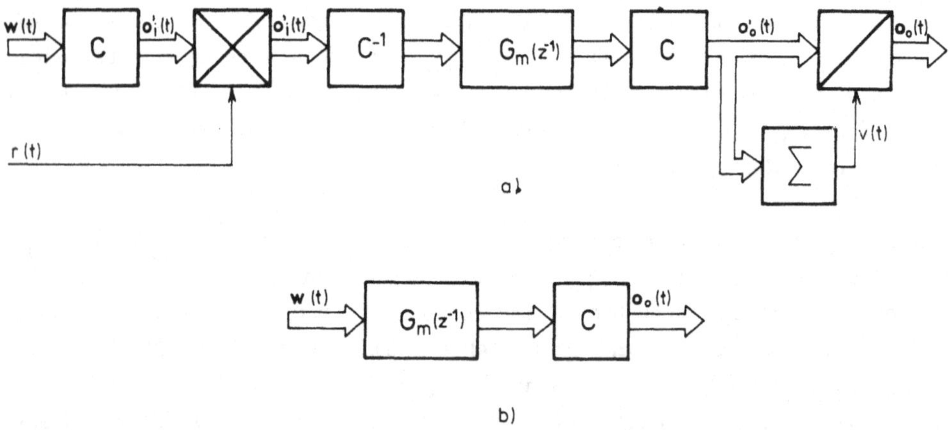

Fig. 8.1-2 a. Dynamic model of a raw mill with respect to the composition variables
b. Simplified dynamic model of a raw mill with respect to the composition variables

Chapter 7. The resultant control engineering model was shown in Fig. 7-5, while the joint model of the raw mill-silo system for batch silos was presented in Fig. 7-8.

It can be concluded from the above models that composition control results in a multivariable coupled system. If the task can be formulated for the oxide variables, then the system may be regarded as linear; if, however, the moduli values are also required, the system becomes nonlinear. It should be mentioned that the parameters of batch silo are time-varying, so the problem is to control a nonlinear time-varying system.

The aim of the control is to charge the silo in such a way as to achieve a desired average silo content and to keep the fluctuations of the components around this reference value as small as possible. The homogenization silos serve these purposes, i.e. for decreasing the inhomogeneities - usually by air injection - and for preventing the discharge of nonhomogeneous material into the rotary kiln. If the fluctuations in the composition of raw meals make it possible, continuous blenders can be also used, which results in a considerable saving in cost. Control systems applied to date have already attained the level predicted by classical control theory.

To survey the possible control systems, let us consider again Fig. 7-5. The model of the raw mill relating to the composition variables can be given by Fig. 8.1-2/a. Here and throughout this chapter t refers to discrete time, and argument z^{-1} to the discrete model [8]. Thus $G_m(z^{-1})$ is the discrete

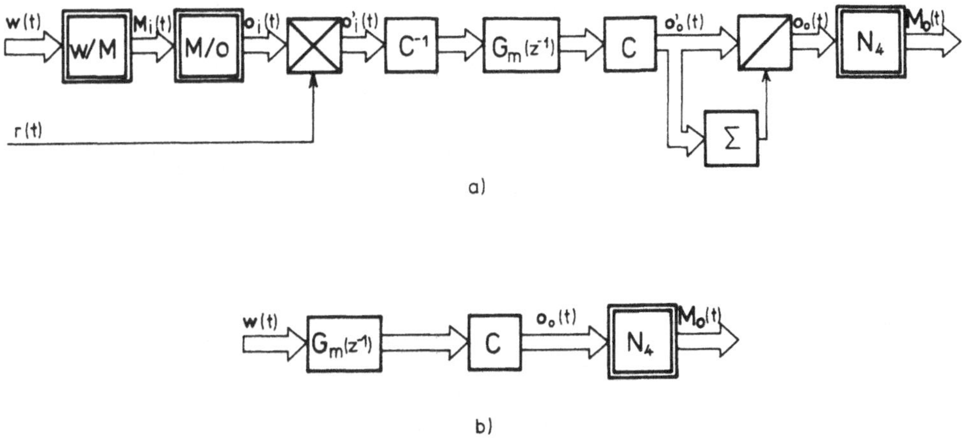

Fig. 8.1-3 a. Composition model of a raw mill containing moduli values
b. Simplified composition model of a raw mill containing moduli values

transfer function of the mill relating to the partial material flows. As long as the sampling time of the composition control is much higher than the transient of the mill relating to the material flows (mostly this is the case when a sampling time of half an hour or longer is applied), we get Fig. 8.1-2/b. The relationship between the belt scales (**w**(t)) and the oxide compositions ($\mathbf{o}_0$(t)) of the mill product is - with a good approximation - a multivariable linear dynamic coupled relationship.

By further reconfiguration of Fig. 8.1-2/a we get Fig. 8.1-3/a where variables having moduli values and the nonlinear elements **w/M**, **M/o**, $\mathbf{N_4}$, necessary for their calculations, are indicated. By the proper choice of sampling time (long enough) we get Fig. 8.1-3/b where the nonlinearity appears at the output of our system. The nonlinearity $\mathbf{N_4}$ refers to the definition equations (7.1),(7.2), (7.3) of the moduli; **M/o** refers to (7.16); while **w/M** results from the joint application of (7.9) and the definition equations. **C** is the composition matrix by (7.6); $\mathbf{G}_m(z^{-1})$ is the discrete equivalent of $\mathbf{Y}_m(s)$, given by (7.18), i.e.:

$$G_m(z^{-1}) = \text{diag} < \ldots, \frac{c_i}{1+d_i z^{-1}}, \ldots > z^{-d} \; ; \qquad (8.1.1)$$

thus it involves linear dynamics without interactions, having first-order lags in its every channel. Due to the relatively long sampling period, this model yields a rather good approximation; furthermore, common delay z^{-d} can be used.

Applying the simplification conditions of Figs. 8.1-2/b and 8.1-3/b the joint model of the mill-silo system is shown in Fig. 8.1-4, obtained similarly to Fig. 7-8. In the Figure $\mathbf{G}_s(z^{-1})$ is the discrete

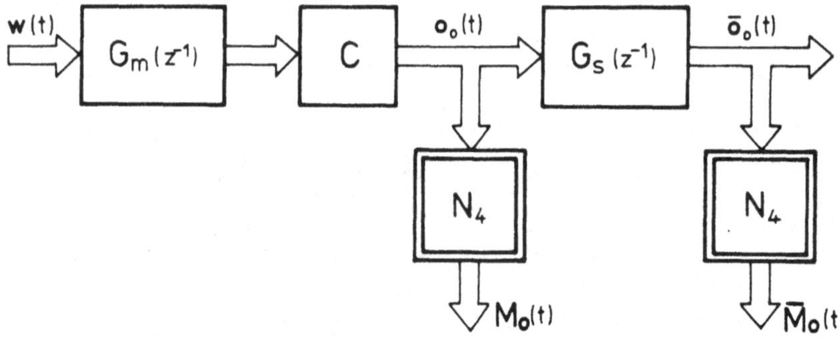

Fig. 8.1-4 Joint, simplified, model of a mill-silo system

model of the silo. For a batch silo [90] with simple calculations we get

$$G_s(z^{-1}) = \frac{b_0}{1+a_1 z^{-1}} \, I \, , \qquad (8.1.2)$$

where b_0, a_1 are time-varying parameters:

$$b_0 = b_0(t) = \frac{v(t)}{q(t)} \qquad (8.1.3)$$

and

$$a_1 = a_1(t) = -\frac{q(t)-v(t)}{q(t)} \, . \qquad (8.1.4)$$

Here

$$q(t) = \sum_{i=1}^{t} v(t) + q(0) \qquad (8.1.5)$$

is the instantaneous silo content. The notations are in accordance with those used in Chapter 7, but now t means discrete time. Fig. 8.1-5 shows the time functions (8.1.4) and (8.1.5) where N indicates the time of silo charging. The dynamics represented by $G_s(z^{-1})$ used to be called *silo-integration.*

The number of degrees of freedom always has a significant role in the choice of control loops. It is

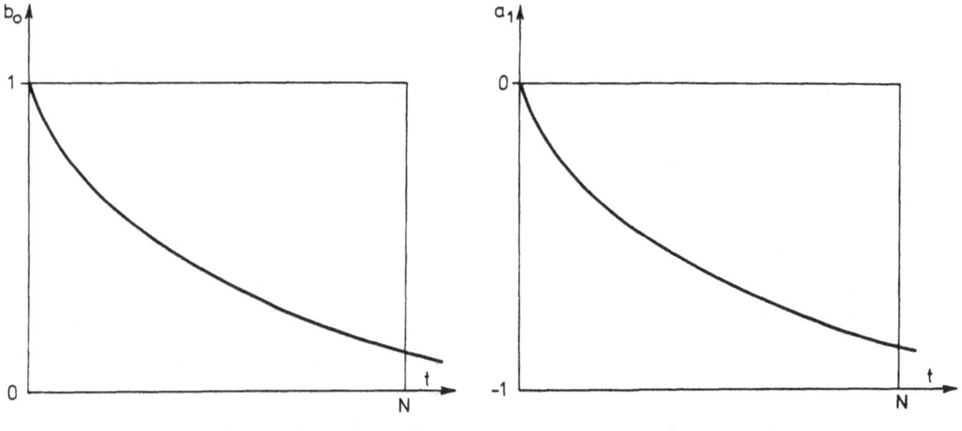

Fig. 8.1-5 Time dependence of the varying parameters of a batch homogenizing silo

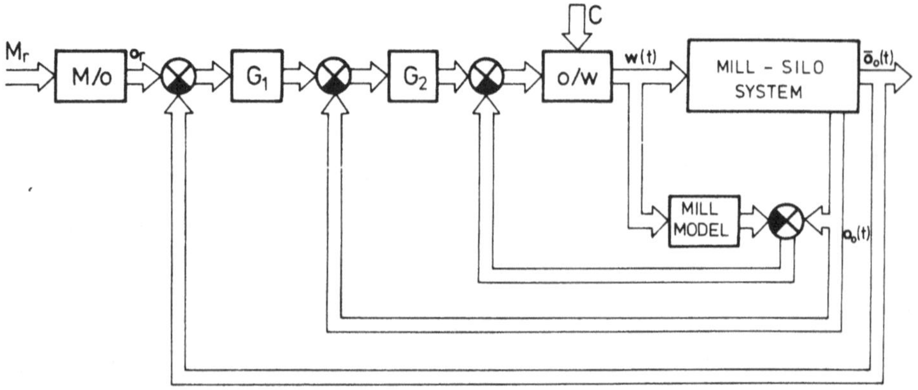

Fig. 8.1-6 Block-scheme of composition control based on oxide values

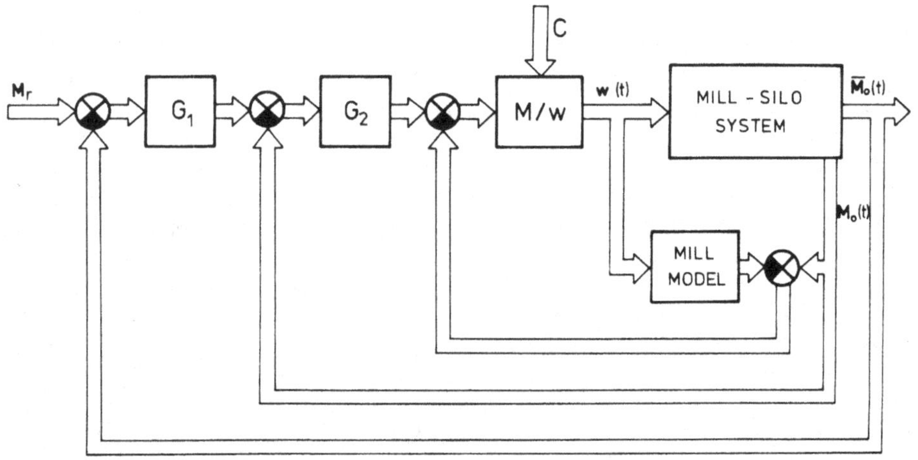

Fig. 8.1-7 Block-scheme of composition control based on moduli values

quite obvious that having k raw meal balance equations, oxide values number k, and (k-1) moduli values, can be adjusted (if the composition matrix makes this adjustment possible at all). Accordingly, the usual control systems follow essentially one of two directions: to control the oxide values or to control the moduli values. Figs 8.1-6 and 8.1-7 show complex control systems for these two approaches. The means of control is the mill model in the inner loop which operates only when the mill behaves differently from its model and provides fast compensation if the model is accurate. The zero value of the steady-state error is ensured by G_2 being a (usually decoupled)

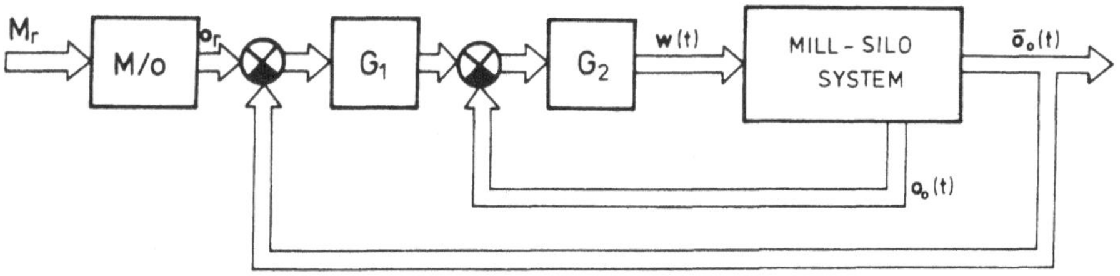

Fig. 8.1-8 Block-scheme of composition control with unknown composition matrix

multi-loop PI regulator. $\mathbf{G}_1$ in the outer loop is also a PI (with constant or time-varying parameters) and it serves for taking the silo integration into consideration. Both approaches have such transformations as **o/w**, or **M/w** which require knowledge of the composition matrix **C** (to be considered as one of the disadvantages of these methods). **C** can be taken either as the average of earlier analyses or can be substituted by the current average composition of the silo. $\mathbf{o}_r$ and $\mathbf{M}_r$ refer to the vectors of the reference signals; the other symbols have been already used in Chapter 7. If, in oxide control, the moduli values are given as reference values, then the block **M/o** of Fig. 8.1-6 is also required.

The three moduli values include four oxides (C, S, A, F); so according to the most frequently used solution, the apportioning of four raw meals is controlled. If more than four raw meals are mixed, the conversion **o/w** (which is actually a matrix inversion) is completed by the calculation of an optimum cost function (i.e. by linear programming). The control of the three moduli requires also four raw meals. In Hungary, in the majority of cases, three raw meals are fed; accordingly, two moduli can be controlled; the third one results.

In moduli control, the number of the actuating signals is greater by one than that of the signals to be controlled, while for oxide control they are equal. In the latter case **w**(t) can be adjusted directly by the regulator $\mathbf{G}_2$. Such a control system is shown in Fig. 8.1-8. The main advantage of this system is that knowledge of the composition matrix is not required (though such knowledge could make control faster and more accurate).

A further disadvantage of the above control schemes - besides the desirability of knowing **C** - lies in the lack of adaptibility. This means that a system whose control parameters have been set to the optimum can show quite different - worse transient - after a specific period if the mill parameters or the composition of the raw meal are changed, or if, in the case of moduli control, the system is operating at a new working point (due to the highly nonlinear relationship between the moduli values).

The mill-silo composition control system is affected by disturbances mainly from two frequency ranges: by quick, stochastic changes and by slow ones regarded as parameter changes. Accordingly, a control system is required which minimizes the effect of stochastic disturbances and tunes its parameters to the actual circumstances. Because of the long dead-time, the application of the predictive control principle seems to be reasonable. For batch silos the average composition control needs further special considerations.

These aims can be achieved by using a multiple-input multiple-output self-tuning minimum variance (MIMO-ST-MV) control based on d-step ahead adaptive prediction [119].

Composition control has to ensure minimum variation of the composition of the raw meal around a prescribed reference value, as well as an average composition for the silo content. These two control requirements can be satisfied by minimizing a general loss function

$$V = E\{\|y(t+d)-y_r\|^2 + \gamma \|y_a(N)-y_r\|^2 |_t\} = \min \qquad (8.1.6)$$

where $y(t+d)$, y_r and $y_a(t)$ are the vectors of the measured composition, the desired (reference) values and the average composition, respectively. The time $t=N$ indicates the end of silo charging, γ is a positive penalty constant, d is the prediction horizon. Here E{...} stands for mathematical expectation. (Note that any $y_a(t+k)$ for $(t+k<N)$ can also be used.)

The minimization of V leads to the so-called MIMO-ST-MV-RAFT [119] control strategy which also ensures the desired average composition during silo charging. The abbreviation RAFT refers just to this fact (Required Average for Finite Time). Optimal control can be achieved by a special cascade closed-loop system, shown in Fig. 8.1-9. In the inner loop there is a MIMO-ST-MV

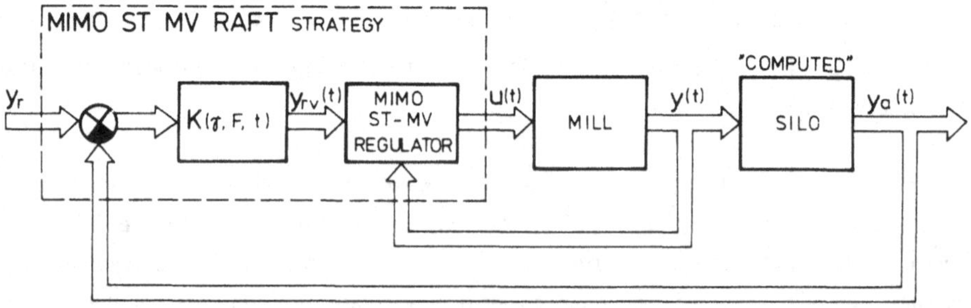

Fig. 8.1-9 MIMO-ST-MV-RAFT adaptive control system of composition control

regulator whose varying reference signal $y_{rv}(t)$ is adjusted by a time-varying P regulator in the outer loop

$$K(\gamma,F,t) = K(\gamma,F,t)\ I \qquad (8.1.7)$$

having the same gain in each channel. The minimization of V is equivalent to the task

$$Q = \|y(t+d|t)-y_r\|^2 + \gamma\ \|y_a(N)-y_r\|^2 = \min \qquad (8.1.8)$$

under certain conditions [119], which leads to the control law

$$y(t+d|t) = y_{rv}(t+d) = y_r + K(\gamma,F,t)[y_r-y_a(t+d|t)], \qquad (8.1.9)$$

shown in Fig. 8.1-9, where the argument t+d|t relates to a d step ahead predicted value. Here

$$y_a(t+d|t) = \frac{y_a(t)q(t) + \sum_{i=t+1}^{t+d-1} v(i)y_{rv}(i)}{q(t+d-1)} \qquad (8.1.10)$$

and

$$K(\gamma,F,t) = \frac{\gamma[\sum_{i=t+1}^{t+d-1} v(i)][F-q(t+d-1)]}{F^2 + [F-q(t+d-1)]^2} \qquad (8.1.11)$$

and F is the full silo content [90].

A rather good approximation for $y_a(t)$ can be obtained by the *silo-integration* $G_s(z^{-1})$; accordingly, $y_a(t)$ contains calculated values.

The above mentioned MIMO-ST-MV-RAFT adaptive regulator was used for composition control by the choice of

$$y(t) = M_o(t) \qquad y_a(t) = \bar{M}_o(t) \ . \qquad (8.1.12)$$

Fig. 8.1-10 shows the intantaneous ($KS_t(t)$, $AM(t)$) and the average ($\overline{KS}_t(t)$, $\overline{AM}(t)$) moduli values

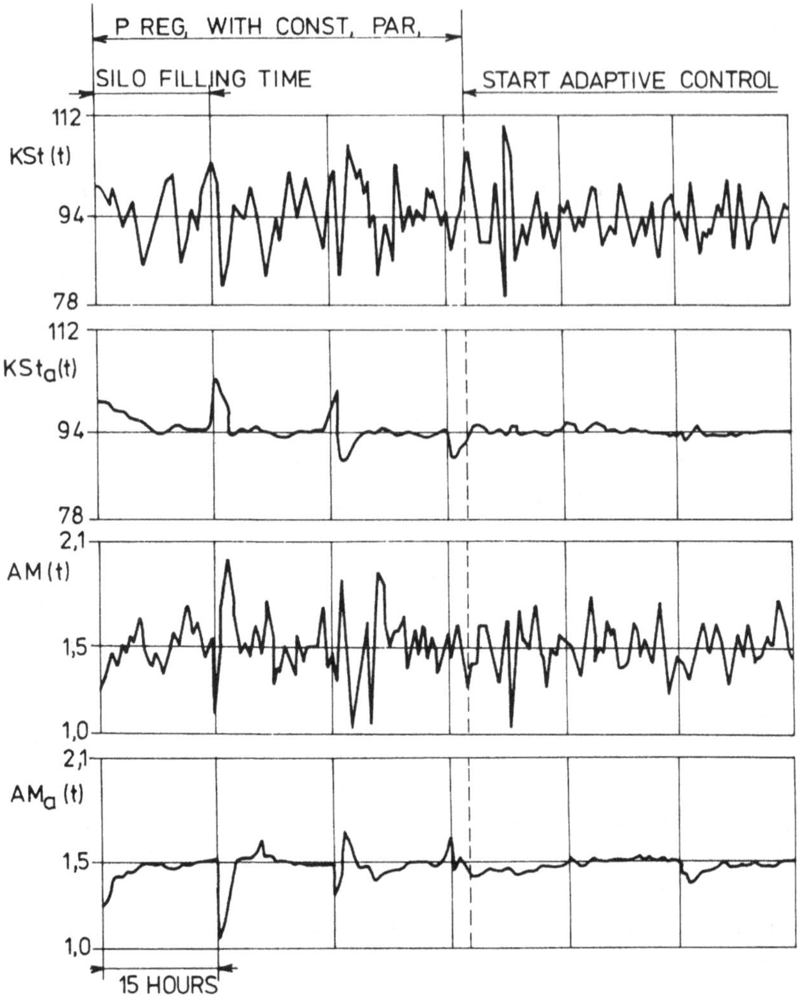

Fig. 8.1-10 Time functions of the controlled variables for MIMO-ST-MV-RAFT regulator

under conventional control, for the first three charging periods, and under adaptive control, for the last three periods. Thus the control was performed for two moduli values using three raw meal balance equations. Two scale-rates were included; the third one was obtained as $w_3=1-w_1-w_2$, so the input signal was

$$u(t) = \tilde{w}(t) = [w_1(t), w_2(t)]^T . \qquad (8.1.13)$$

Multivariable regulators usually have many parameters to be adjusted; therefore, special solutions which decrease drastically the number of parameters to be adapted have especially great importance. An example will be given showing this for a special form of the composition matrix C. In our case C had the following form (to a good approximation):

$$C \cong \begin{pmatrix} C_1 & 0 & 0 \\ 0 & S_2 & 0 \\ 0 & A_2 & 0 \\ 0 & 0 & F_3 \end{pmatrix} \tag{8.1.14}$$

This structure ensures that the following relationships are valid to a good approximation for the most important oxides

$$C \cong C_1 w_1 \quad ; \quad S \cong S_2 w_2$$
$$A \cong A_2 w_2 \quad ; \quad F \cong F_3 w_3 \quad . \tag{8.1.15}$$

In this case the moduli values can be calculated in a very simple manner:

$$KS_t = \frac{100\ C_1}{2.8 S_2 + 1.1 A_2} \left(\frac{w_1}{w_2} \right) \tag{8.1.16}$$

$$AM \cong \frac{A_2}{F_3} \left(\frac{w_2}{w_3} \right) \quad \rightarrow \quad \frac{1}{AM} = \frac{F_3}{A_2} \left(\frac{w_3}{w_2} \right) \tag{8.1.17}$$

$$SM \cong \frac{S_2 w_2}{A_2 w_2 + F_3 w_3} \rightarrow \frac{1}{SM} = \frac{A_2}{S_2} + \frac{F_3}{S_2} \left(\frac{w_3}{w_2} \right) \quad . \tag{8.1.18}$$

It is seen from the above that the multivariable system, originally nonlinear with respect to the moduli values, can be linearized and decoupled by the choice of

$$u_1 = \frac{w_1}{w_2}$$
$$u_2 = \frac{w_3}{w_2} \tag{8.1.19}$$

It seems to be reasonable to use the reciprocal values of the moduli AM and SM. In our case

$$y_1(t) = \frac{KS_t(t) - KS_{tr}}{KS_{tr}} \qquad (8.1.20)$$

and

$$y_2(t) = \frac{\dfrac{1}{AM(t)} - \dfrac{1}{AM_r}}{\dfrac{1}{AM_r}} \qquad (8.1.21)$$

are chosen [90].

By decoupling we got the two **S**ingle **I**nput-**S**ingle **O**utput (SISO) adaptive control loops shown in Fig. 8.1-11. In the outer loop $G_1(z^{-1})$ can be substituted by a PI regulator or a parameter varying regulator

$$G_1(z^{-1}) = \frac{p_o(1+a_1 z^{-1})}{1-z^{-1}} \qquad (8.1.22)$$

where

$$p_o = \frac{1}{b_0(2d-1)} = \frac{q(t)}{v(t)(2d-1)} \qquad (8.1.23)$$

and

$$p_1 = p_o a_1 = -\frac{q(t)-v(t)}{v(t)(2d-1)} \qquad (8.1.24)$$

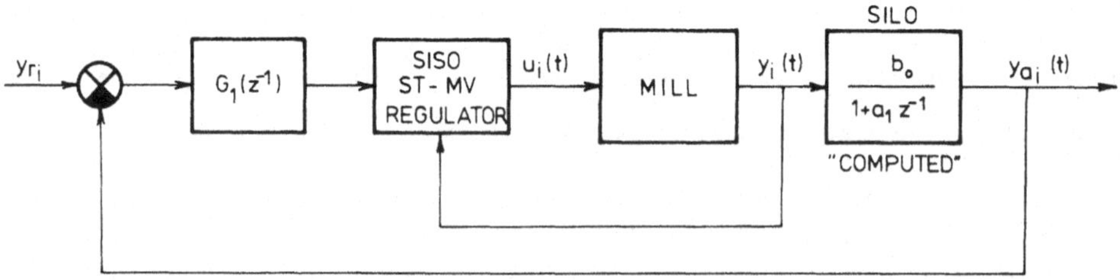

Fig. 8.1-11 Block-scheme of decoupled SISO-ST-MV composition control

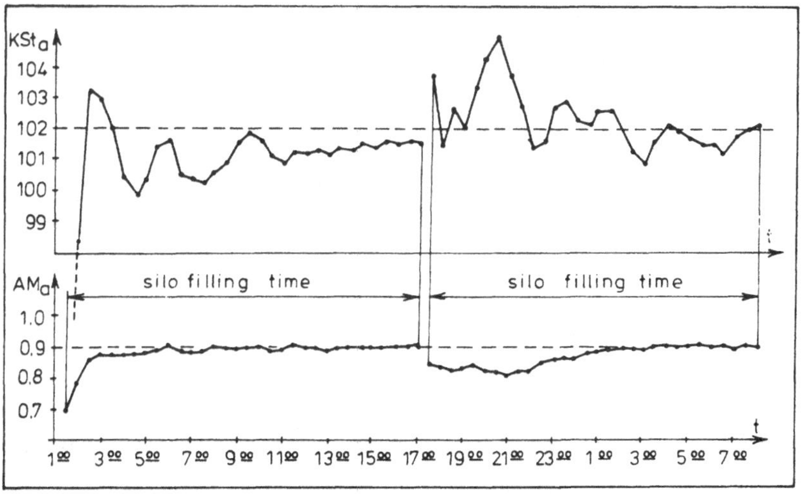

Fig. 8.1-12 Time functions of the controlled variables for SISO-ST-MV regulator

Fig. 8.1-12 shows $\overline{KS}_t(t)$ and $\overline{AM}(t)$ for the period of silo charging. While the previous MIMO regulator required that 20 parameters be adapted, here only 8 had to be updated in the two regulators. If natural decoupling resulting from the structure of **C** is not possible, an adaptive control should be used, where the adaptive identification of the process is directed to learning only the static relationships, and dynamic control is performed by a regulator with constant parameters. Such a system is given for oxide variables in Fig. 8.1-13. Here $G_2(z^{-1})$ is a PI regulator, while

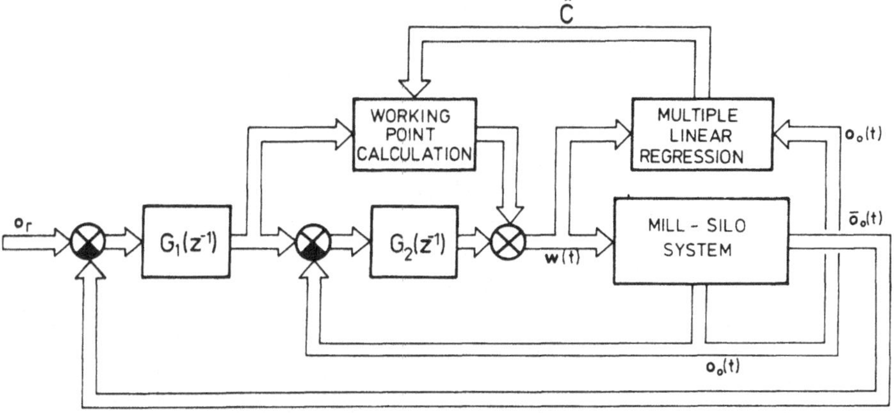

Fig. 8.1-13 Composition control on the basis of adaptive identification of static relationships

$G_1(z^{-1})$ is a PI or $\mathbf{K}(\gamma,F,t)$ regulator. The adaptive estimation of $\hat{C}$ can be obtained by the recursive form of multiple linear regression [116]. The errors resulting from the inaccurate calculation of the working point are made equal to zero by $G_2(z^{-1})$.

According to our present knowledge, such two-level hierarchic adaptive regulators can be considered as the application of the most advanced control technique [123].

8.2. Control of the quantity produced

The composition control system discussed in the previous section can be found at the front-end of cement producing technology, i.e. at the raw mills. The necessity for control of the quantity produced (and its fineness) to be discussed in this section (and in the next) arises at the end of the technology, i.e. in connection with cement mills, though the results obtained here can be applied to the raw-mills, too; but the requirement of the latter for this type of control is not significant. So let us consider this section as a survey of control systems for ensuring maximum ground material in closed-circuit mills.

As is well known for closed-circuit mills, the main problem of control is that changes caused by the recycling grit may give rise to material waves, which can lead to clogging of the mill, if the fresh feed is not properly apportioned. Therefore, the primary aim of most control systems is to keep the mill load at a constant level. The simplest way of achieving this is by a so-called *input-regulator,* shown in Fig. 8.2-1, where x_{max} and r_{max} are the permitted maxima of the total feed and fresh

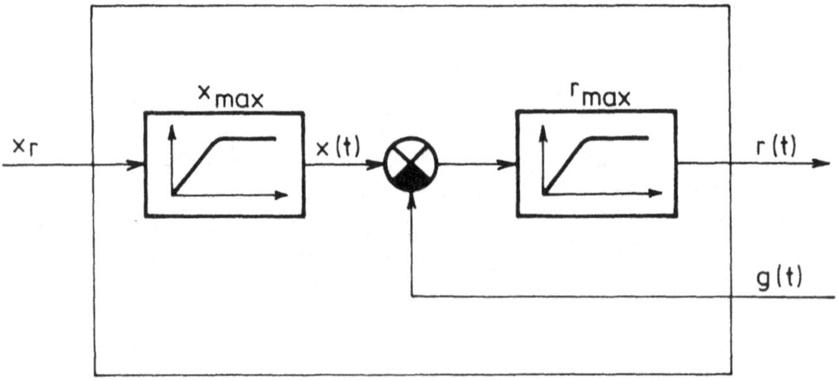

Fig. 8.2-1 Block-scheme of the input regulator

feed, respectively. If neither of these limits is reached, the regulator ensures the desired total feed x_r by adding the necessary quantity of fresh feed to the actual value g(t) of the grit. (The same notations and nomenclature are used here as in the previous chapters.)

Though the input regulator keeps x(t) at a constant value, it is not able, in itself, to guarantee the smooth operation of the mill. m(t) changes with different rates r/g even if x(t) is constant, due to the varying residence time caused by the different size distributions. (See related models in Chapter 5.) As a consequence, stable operation of the mill can be ensured only by feedback (the input regulators have no feedback!).

Closed-circuit grinding is a complicated nonlinear system with large dead-times and varying time constants, where possibilities for intervention are very limited. Essentially, only the fresh feed and the r.p.m. of the spreading plate of the classifier may be chosen as variables for intervention. (The air-flow of the mill is usually kept at a constant value for safety reasons; similarly, the r.p.m. is set to a value near the optimum.)

The most advantageous case is when the three main material flows: fresh feed, grit (or their sum, the total feed), and the final product are measured; and, very rarely, the mill product, too. Usually the power consumption of the elevator, which is a very noisy signal, is measured; it is proportional to the mill product. The electric ears arranged along the mill also yield rather noisy signals, but by appropriate filtering we get signals proportional to the filling factors along the mill length.

The main sources of disturbances are the size distribution of the clinker and the changes in grindability.

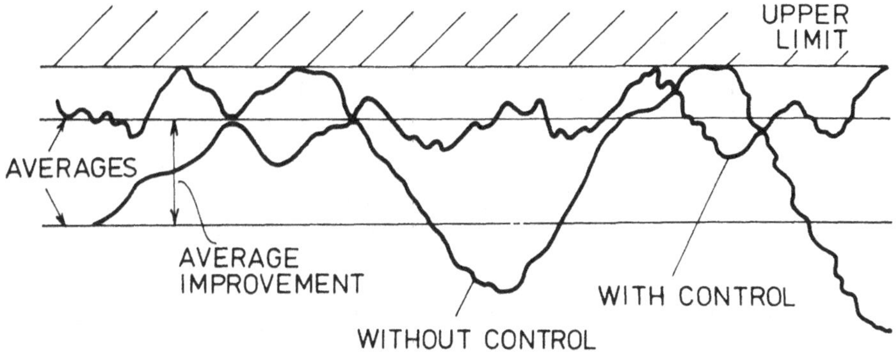

Fig. 8.2-2 The mill product using two-step quantity control strategy

The energy consumption of grinding is very high; the majority becomes heat and only a small percentage is used for breakage. The consumption of the mill depends hardly at all on the load, so minimal specific energy consumption can be achieved by maximization of production. By increasing the fresh feed, as has been seen in Chapter 5, the final product must increase due to the law of indestructibility of matter. We have seen that there is a stability limit for the fresh feed, which ought to be approached as nearly as possible. (At this point the mill product has maximum with respect to the filling factor.) Theoretically it is possible to develop a method which keeps the system at this optimum point [50], [121]. In practice, however, this point lies beyond the mechanically permissible load, so it cannot be attained. In the majority of cases the drive of the elevator (the clutch) cannot take a higher load.

When the object of control is to keep the controlled value as close as possible to an upper limit, a two-stage procedure seems to be the most effective strategy, where in the first step the variance of the variable is minimized, then in the second step, by changing the reference signal, the upper limit is approached to the extent permitted by the control error. The success of this control strategy is illustrated in Fig. 8.2-2, i.e. the best strategy is the one that ensures the smooth operation (without swinging) of the mill and increases the fresh feed step-by-step to the possible limit.

Most known and applied control systems have single or multiple cascade loops. The main difficulty lies in the fact that essentially the fresh feed represents the only intervening variable. Further output variables, e.g. the filling factor, can be used only as inner measured signals to make the system faster.

Fig. 8.2-3 gives a survey of different systems applied in practice.

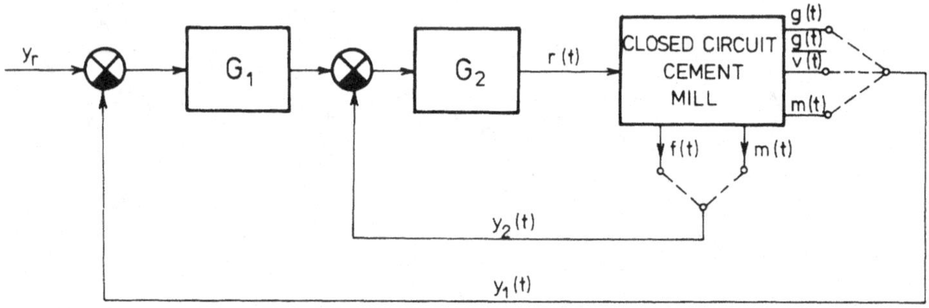

Fig. 8.2-3 Survey of quantity control methods widely used for closed circuit cement mills

In the outer loop the following signals can be chosen for control:

- mill product: m(t) (or the elevator power consumption, which is proportional to this),
- grit: g(t),
- recycling rate : g(t)/v(t) (or any other equivalent quantity).

In the inner loop the controlled signal $y_2(t)$, belonging to regulator G_2, may be chosen as:

- filling factor: f(t) (or the signals of the electrical ears, which are proportional to this),
- mill product: m(t), (if $y_1(t)=g(t)/v(t)$).

The setting of the reference value y_r depends on the choice of $y_1(t)$; usually its value is determined empirically by the operator. The advantage of choice $y_1(t)=m(t)$ is that in this case y_r is proportional to the production, so the upper limit of the capacity can be achieved step by step.

It can be stated from the above that g(t) is usually measured directly, but the same is not true for m(t). It is a very rare case when g(t) and v(t) are measured simultaneously, so that g(t)/v(t) becomes known.

As we have seen in previous chapters, closed-circuit grinding represents a multivariable, nonlinear, distributed parameter process having feedback and dead-times, whose lumped parameter approach is dependent on the load.

It has been mentioned that the changes in the size distribution and grindability of the clinker always cause stochastic disturbances in the mill. This fact requires the application of advanced algorithms along with classical control methods. The choice of the optimal point corresponding to the extremum static characteristics and the closest approximation of the boundary conditions need also an advanced control engineering background. According to our present knowledge, the best solution is obtained by a two-level, hierarchic adaptive control system whose block-scheme is shown in Fig. 8.2-4. Compared with Fig. 8.2-3, this system is completed by a supervisory adaptive reference signal control. The inner loop contains a SISO-ST regulator [9] which allows optimal compensation of stochastic disturbances of higher frequency; it is able to follow parameter changes due to varying load.

In the outer control loop the task of the PI regulator is to maintain the designated output variable at the reference signal received from the supervisory control.

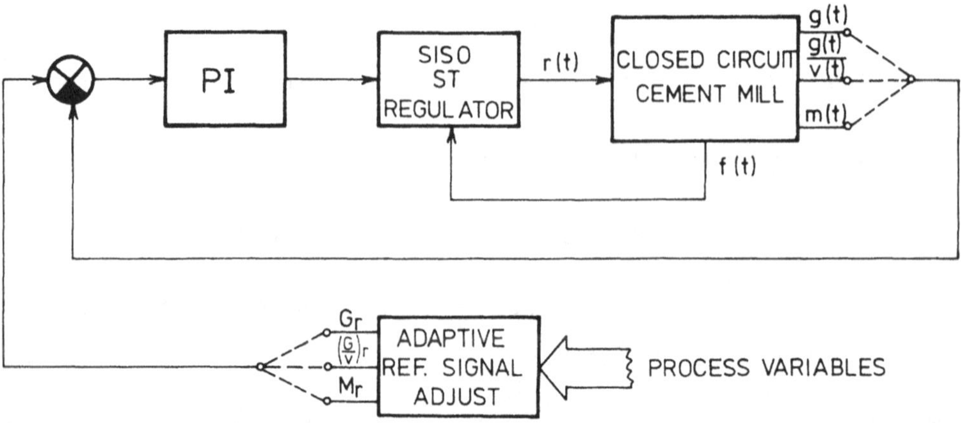

Fig. 8.2-4 Two-level, hierarchic adaptive quantity control of closed-circuit cement mills

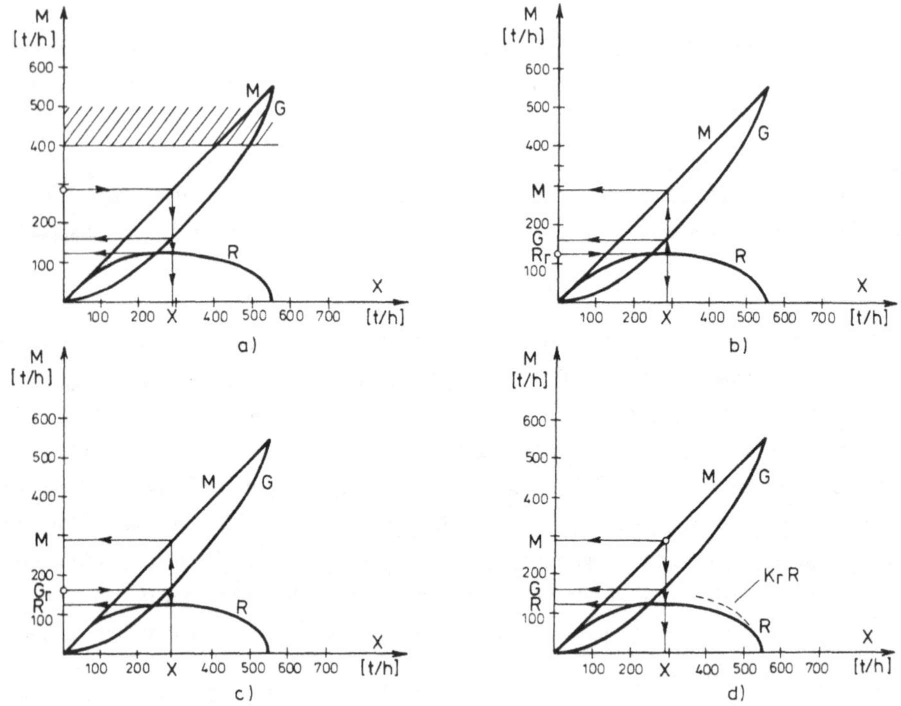

Fig. 8.2-5 Formation of the working point of closed-circuit cement mills for different basic control cases

Fig. 8.2-5 shows the static characteristics (M(X), G(X), R(X)) of a closed-circuit cement mill and the position of the working point in typical situations. In a, b, c the reference values M_r, R_r, G_r are required, respectively. In d, the recycling rate K=X/R=M/V is prescribed, and all the other parameters are obtained accordingly. b represents open-loop control; a, c and d represent closed-loop control. In d, the relationship between the required recycling rate K_r and reference value G/V_r prescribed for the ratio g(t)/v(t) is given by

$$K_r = 1 + \left[\frac{G}{V}\right]_r = 1 + \left[\frac{G}{R}\right]_r \qquad (8.2.1)$$

It is evident that in this case the only possible region for the reference value is

$$\frac{1}{\left.\dfrac{\partial R}{\partial X}\right|_{X=0}} < K_r < 1 + \frac{G_o}{R_o} \quad . \qquad (8.2.2)$$

(For lower K_r, there is no intersection between K_rR and M(X), while K_r over this limit would lead to an unstable working point.)

It can be easily seen that - due to the restriction $M < M_{max}$, which almost always obtains - in *a* this limit has to be approached through the reference value M_r. A brief survey of the strategies applied in supervisory adaptive control follows.

The hierarchically supreme control-loop determines the reference signal for the subordinate system, on the basis of the static characteristic of the mill.

In the simplest and most practical case this system follows the variation of m(t) and, computing the variance of this signal (σ_m) for a given interval, it calculates the reference signal M_r

$$M_r = M_{max} - 3\,\sigma_m \qquad (8.2.3)$$

taking a specified safety region (e.g. $3\sigma_m$) into consideration. In this way adaptive control can maximize production under restrictions. Where the above restriction does not exist, different methods of extremum control can be applied [7], [60], [117], [121], but do not forget that only the

Fig. 8.2-6 Extremum control based on relationships between the material in the closed-circuit grinding system and the final product

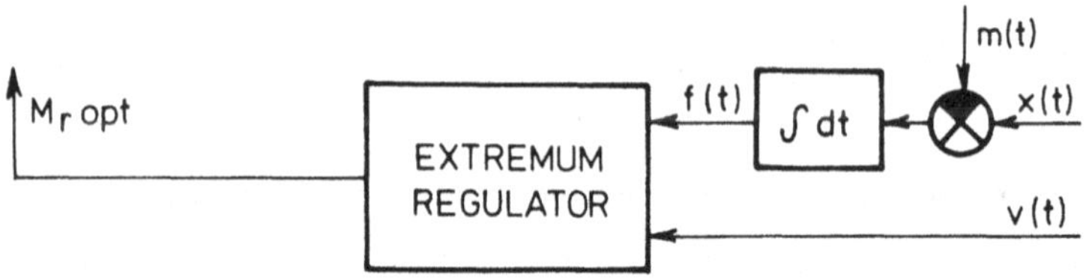

Fig. 8.2-7 Extremum control based on the relationships between the material in the mill and the final product

Fig. 8.2-8 Extremum control on the basis of the mill product and final product

region on the left of the extremum point yields stable working points, so the optimum at the same time represents the limit of instability.

One of the principles of mill extremum control known from the earliest time is shown in Fig. 8.2-6 [202] which is based on the simplest qualitative mill model shown in Fig. 5.1-2, where the filling factor is calculated as the integral of the difference between the fresh feed and the final product. Extremum control may operate by using correlation or regression calculations between the filling factor and the final product (see later). Note here that f(t) means the material present in the whole closed-circuit system.

Fig. 8.2-7 shows a modified version of the above system where, according to the complete mill model given in Fig. 5.1-5, the filling factor is calculated by integrating the difference between the total feed and the mill product. Here f(t) represents the material in the mill.

It was mentioned in the discussion of qualitative mill models that the mill product is proportional to the filling factor over a wide range. The solution given in Fig. 8.2-8 is based on this fact; the extremum regulator uses m(t) and v(t) proportional to f(t) to search for the extremum.

Do not forget that the properly filtered signal of the electric ear may be also proportional to the filling factor.

Applied extremum regulators operate mostly according to Fig. 8.2-9/a. Using the correlation technique of linear regression, a signal proportional to the gradient dV/dF is generated from v(t) and f(t), and an intervention is performed by an integrator in the direction determined by the sign of the derivative. The condition of persistent excitation is ensured by applying a low frequency square-wave signal, of amplitude of 5 to 10 percent. of working point value, to the input. This method requires the artificial perturbation of the system and a relatively long integrating time for the actuator. Although it is not illustrated in the figure, time delays have to be applied, especially in computer control, to obtain coherent data for processing (relating to the same times). So the effect of pure time delays (dead-times) can be partly compensated for.

Especially in computer applications, the recursive least-squares method [211] for fitting a linear model has proved to be better than the correlation method, where the coefficient of the linear term gives the gradient (i.e. it is proportional to the correlation coefficient). A more advanced application of the regression technique is the quadratic approximation, when the optimum is obtained as the extremum of a parabolic model. Fig. 8.2-9/b refers to this method.

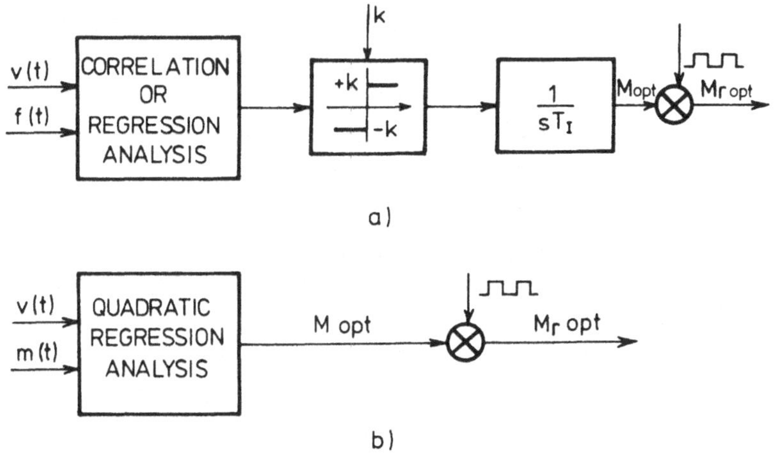

Fig. 8.2-9 a. and b. Schema of the extremum control systems applied in the practice

As the inner regulators are adjusting m(t) on the basis of M_r with a speed determined by the mill dynamics, the whole system has nonlinear dynamic behaviour toward $M_r(t)$ and v(t). Consequently, the position of the extremum M_{ropt} can be estimated by a nonlinear, dynamic, (a suitably generalized **Hammerstein**) model fitted to the signals $M_r(t)$ and v(t).

Based on the above principle, a control loop having constant x(t) can be also formulated, and the **Hammerstein** model established for $X_r(t)$ and v(t) can be used to determine X_{ropt}. This approach can be easily adapted to the case when only g(t) is measured and v(t) is not. In this case, the **Hammerstein** model has to be formulated for $X_r(t)$ and g(t), and X_{ropt} can be calculated from the condition dG/dX=1.

All these approaches can be extended by minimum variance control techniques [50].

If the fresh feed r(t) is the reference signal, the dynamics of the closed-loop are determined by the nonlinearity N_2 (see Chapter 5.1). Experience has shown that the target of optimum control has to be the determination of the extremum of N_2 rather than the instability limit corresponding to the extremum of N_1. At the extremum point of N_2 the time constant T_2 becomes infinite; thus $\lambda_2=1/T_2=\lambda_2(F)$ becomes zero at the optimal filling factor F_0. A similar phenomenon may be observed for function $\lambda_2(R)$.

Though the functions $\lambda_2(F)$ and $\lambda_2(R)$ are nonlinear, they could well be used to estimate the extremum, especially in the vicinity of the optimum. In reality it may happen that the single time constant does not provide a sufficiently accurate description (think of the disregard of dead-times); therefore the extremum has to be sought from the reciprocal value of the largest time constant of the transfer function.

If a linear first order dynamic model with signal dependent parameters is identified between v(t) and r(t), or v(t) and f(t), the position of the extremum can be also estimated from the signal dependence assumed to be linear for the time constant [123].

8.3. Fineness control

Nowadays, control of the fineness and specific surface of cement becomes a real requirement which has to be met by the building material industry. In spite of this, apart from the very few references in the literature (relating to wet grinding only), to our knowledge there is no such control system in use in industry. Experimentation continues. The main difficulty lies in the continuous, or at least automated, batch perception. A problem of less importance is the large extent of uncertainty in modelling closed-circuit grinding. In this regard, we attach great importance to the direction indicated by the complete dynamic size-distribution model in Chapters 5.2 and 6.

(Here we do not intend to deal with the problem that the current laboratory technique usually measures the so-called **BLAINE** value instead of the specific surface, because the relationship for small changes is proportional and the requirements can be transformed to **BLAINE** values.)

It was stated in Chapter 6 that the specific surface of the final product is affected by the material flow conditions (mostly by the load of the air-classifier, i.e. by the mill product), on the one hand, and by the r.p.m. of the spreading-plate, on the other. One of the possible control strategies uses this latter quantity for intervention as its effect may be observed faster. The material flows are then variables of the production control.

These considerations also confirm that a strong interaction exists between the control processes directed to the maximalization of the final product and to the specific surface, i.e. fineness control of the ground material.

Fig. 8.3-1 represents a joint adaptive quantity and fineness control loop, using Fig. 8.2-4. The quantity control is directed to the mill product through reference signal M_r. As is usual in industrial practice, the specific surface s(t) is measured by laboratory analysis over long time intervals, so it is appropriate to apply a slow I regulator in this loop which intervenes in the process via the r.p.m. c(t) of the classifier. Obviously, in this loop, the ST regulator can also be applied to improve the dynamic behaviour. It is advisable to make a feedforward control from c(t) - for the purpose of noise compensation - to the inner loop of the quantity control. In this way the material waves resulting from a sudden change of c(t) can be reduced drastically. Based on the consumer's requirements, the reference signal S_r is known in advance for a given period of production. Fig. 8.3-2 shows the operation of a self-tuning regulator with a sampling time of h=15 min. The part *a* of the figure illustrates the learning period, while part *b* represents the stationary operation for a different reference value.

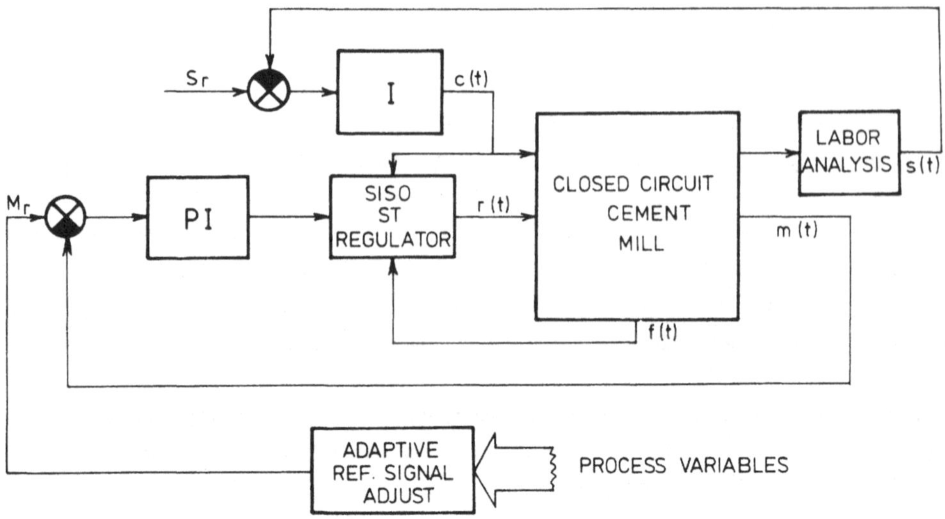

Fig. 8.3-1 Block-scheme of joint quantity and fineness control

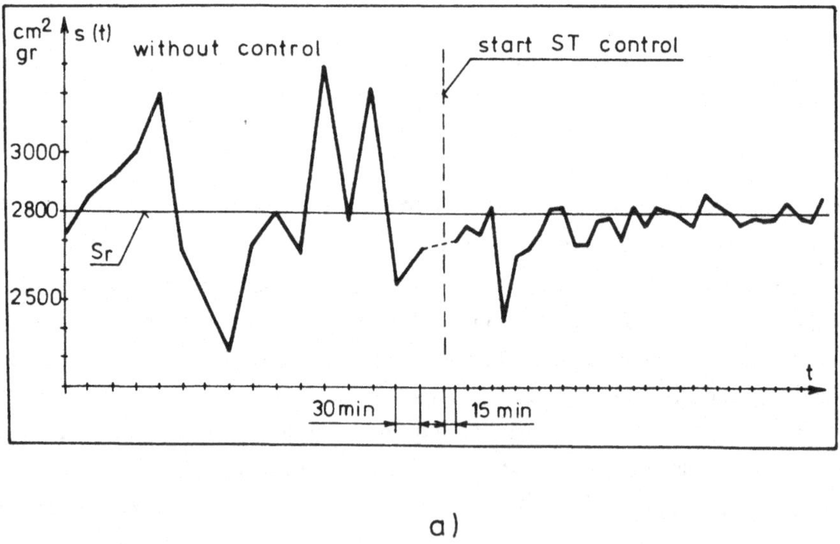

Fig. 8.3-2/a Time function of the **BLAINE** value in the learning period in self-tuning fineness control

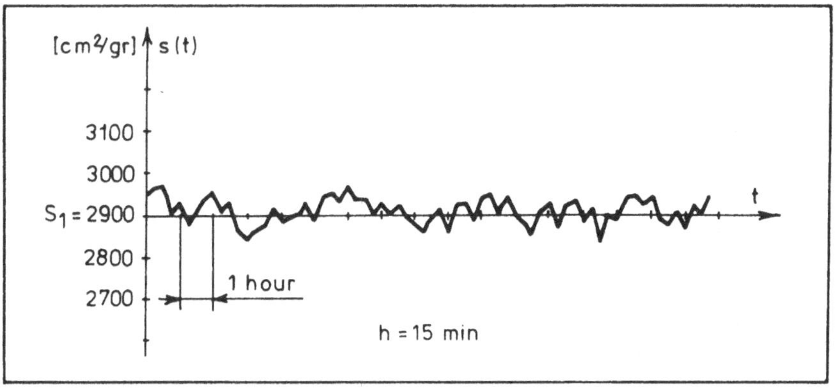

b)

Fig. 8.3-2/b Stationary time function of the **BLAINE** value in self-tuning fineness control

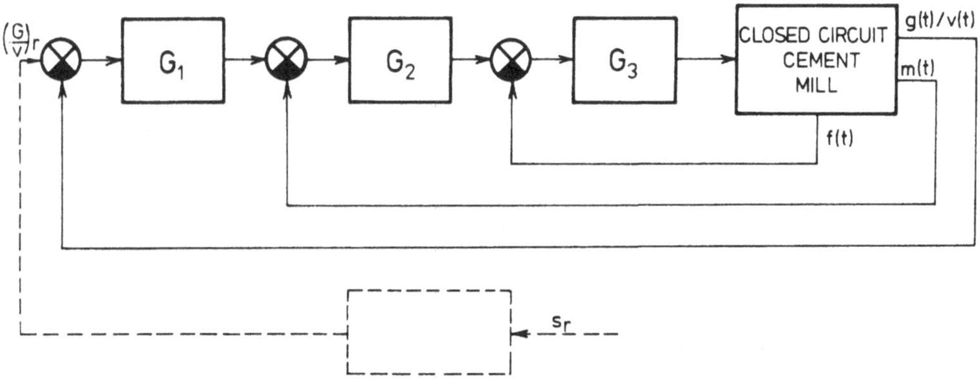

Fig. 8.3-3 Block-scheme of the **BLAINE** value control by keeping the recycling rate at constant value

If we suppose that a given value of the specific surface is a function of the recycling factor, then the control requirement to keep the ratio g/v at a certain value may simultaneously produce a nearly constant **BLAINE** value [61]. Such a control loop is given in Fig. 8.3-3. The main difficulty lies in finding the reference signal $(G/V)_r$ belonging to the required S_r. Unfortunately, this system entails the measurement of too many variables and, in fact, it does not contain feedback from the measured s(t).

Summary of Chapter 8

This chapter treats control problems of grinding where the most advanced results of control theory can be applied directly. Chemical composition - the production - and fineness controls are considered. These control processes are in different units of the cement producing technology: e.g. composition control at raw mills, production control at cement mills. There is already a need to unite these control systems and the time is not far ahead when simultaneous control processes for any closed-circuit grinding mill will be achievable.

We have not discussed control strategies occurring in practice and in this respect we refer to the excellent surveys of the literature [123]. But the most widely adopted solutions are presented here.

From the latest results of advanced control theory, adaptive optimal control algorithms are emphasized, and these could be applied now, using microcomputer techniques.

Symbol nomenclature of Chapter 8

$\mathbf{G_m}(z^{-1})$	discrete transfer function of the mill with respect to the partial material flows
z	variable of z-transform
$\mathbf{o}_i(t)$	vector of oxide composition of the input flow
$\mathbf{w}(t)$	vector of oxide composition of the material flow
c_i	coefficients
d_i	coefficients
z^{-d}	discrete dead-time
w/M, M/o, N_4	nonlinear elements used in the calculation of variables containing moduli values
C	composition matrix of the raw meal
b_0, a_1	time varying parameters of the discrete model
$v(t)$	material flow fed to the silo from the mill [t/h]
$q(t)$	instantaneous material quantity present in the silo
PI	proportional-integral regulator
I	integral regulator
$\mathbf{o_r}$, $\mathbf{M_r}$	reference signals with respect to oxide or moduli values
V	cost or loss function

$E\{...\}$	expectation value
$\mathbf{y}(t+d)$	measured composition vector
y_r	vector of reference values
$t=N$	time of the end of silo filling
$y_a(t)$	vector of average compositions
F	the whole silo capacity [t]
$\mathbf{K}(\nu,F,t)$	matrix of time varying, proportional regulator
w_i	weight scales
$\mathbf{u}(t)$	vector of input signal
$\hat{C}$	estimated composition matrix
r(t), R	fresh feed [t/h]
g(t), G	grit [t/h]
m(t), M	mill product [t/h]
v(t), V	final product [t/h]
f(t), F	filling factor [t]
y_r	reference value of the control
K_r	reference value of the recycling rate
σ	variance
λ_1, λ_2	auxiliary variables

APPENDIX

A1.

Based on the definition formula (2.12) of specific surface we get [154]

$$S = \frac{6}{\gamma}\int_{d_{min}}^{d_{max}} \frac{D'(d)}{d}\, d(d) = \frac{6m}{\gamma d_{max}}\int_{d_{min}}^{d_{max}} \frac{1}{d}\left(\frac{d}{d_{max}}\right)^{m-1} d(d) =$$

$$= \frac{6m}{\gamma (d_{max})^m}\int_{d_{min}}^{d_{max}} d^{m-2}\, d(d) = \frac{6}{\gamma (d_{max})^m}\,\frac{m}{m-1}\left[\, d^{m-1} \right]_{d_{min}}^{d_{max}} =$$

$$= \frac{6}{\gamma (d_{max})^m}\,\frac{m}{m-1}\left[\,(d_{max})^{m-1} - (d_{min})^{m-1}\right] \qquad (A.1.1)$$

where the density function

$$D'(d) = \frac{m}{d_{max}}\left(\frac{d}{d_{max}}\right)^{m-1} \qquad (A.1.2)$$

is used , obtained from Eq. (2.2) of the **GGS** distribution by the choice $d_o = d_{max}$. Where fractions are in geometric progression

$$d_i = a^i\, d_o \qquad (A.1.3)$$

where

$$d_o = d_{max} \qquad (A.1.4)$$

and since a<1

$$d_n = a^n\, d_o = d_{min} \qquad (A.1.5)$$

Using these expressions (A.1.1) can be further developed

$$S = \frac{6}{\gamma d_0^m} \frac{m}{m-1} [(d_0)^{m-1} - (a^n d_0)^{m-1}] = \frac{6}{\gamma d_0} \frac{m}{m-1} [1-a^{n(m-1)}] =$$

$$= \frac{6}{\gamma d_{max}} \frac{m}{m-1} [1-a^{n(m-1)}] \qquad (A.1.6)$$

A2.

It is usually assumed in theoretical investigations that particles of the i-th fraction are of the size d_i. As a matter of fact, their size is ∂, where

$$d_i \leq \delta \leq d_{i-1} \qquad (A.2.1)$$

and, in addition, they have a given distribution, namely:

$$f_i = \int_{d_i}^{d_{i-1}} \frac{\partial f_i}{\partial \delta} d\delta = \int_{d_i}^{d_{i-1}} \frac{f(\delta)}{d_{i-1}-d_i} d\delta = \int_{d_i}^{d_{i-1}} \frac{f(\delta)}{\Delta_i} d\delta \ , \qquad (A.2.2)$$

where essentially the average weight fraction of the i-th interval is calculated from the original distribution [154].

Let $k(\delta)$ be the breakage rate, then the material in the i-th fraction crushed during an infinitely short time dt is

$$k_i f_i \ dt = \left(\int_{d_i}^{d_{i-1}} \frac{f_i(\delta)}{\Delta_i} k(\delta) \ d(\delta) \right) dt . \qquad (A.2.3)$$

Using the formula (3.20) of the comminution function valid for the **GGS** distribution, the material getting to the (i+j)-th fraction from the i-th one is obtained as

$$k_i f_i b_j = \int_{d_{i+j}}^{d_{i+j-1}} \int_{d_i}^{d_{i-1}} \frac{f_i(\delta)}{\Delta_i} k(\delta) m \frac{\delta_b^{m-1}}{\delta^m} d\delta_b \ d\delta =$$

$$= m \int_{d_{i+j}}^{d_{i+j-1}} \delta_b^{m-1} \ d\,\delta_b \int_{d_i}^{d_{i-1}} \frac{f_i(\delta)}{\Delta_i} k(\delta) \frac{1}{\delta^m} d\delta \ . \qquad (A.2.4)$$

If d_0 is defined to be the largest particle size and a is chosen as the sieve ratio, then the limiting points of the intervals (windows) can be obtained as

$$d_i = d_o a^i . \qquad (A.2.5)$$

Using the notation

$$Z_{i,m} = \int_{d_i}^{d_{i-1}} \frac{f_i(\delta)}{\Delta_i} k(\delta) \frac{1}{\delta^m} d\delta \qquad (A.2.6)$$

for (A.2.4), (A.2.3) becomes

$$k_i f_i = Z_{i,o} \qquad (A.2.7)$$

so (A.2.4) can be simplified as

$$Z_{i,o} b_j = m \int_{d_o a^{i+j}}^{d_o a^{i+j-1}} \delta_b^{m-1} d\delta_b Z_{i,m} \qquad (A.2.8)$$

whence

$$b_j = \frac{Z_{i,m}}{Z_{i,o}} (d_o a^{i-1})^m a^{jm} (1-a^m) . \qquad (A.2.9)$$

According to the definition of the comminution matrix, the equality

$$\sum_{j=0}^{\infty} b_j = 1 \qquad (A.2.10)$$

must hold true. Comparing (A.2.9) with (A.2.10), we get

$$\frac{Z_{i,m}}{Z_{i,o}} [d_o a^{i-1}]^m = 1 \qquad (A.2.11)$$

and

$$b_j = a^{jm} (1-a^m) . \qquad (A.2.12)$$

From the latter equation

$$b_o = 1-a^m \ . \qquad (A.2.13)$$

As could be expected, $b_0>0$, if $a<1$ and $m \neq 0$. From (A.2.12) and (A.2.13) we obtain

$$b_j = b_o \, a^{jm} \ . \qquad (A.2.14)$$

Using (3.16) and (A.2.14) we get the expression

$$b_i' = \sum_{j=i}^{\infty} b_j = b_o \frac{1}{1-a^m} - \frac{1-a^{(i-1)m}}{1-a^m} = a^{im} \ . \qquad (A.2.15)$$

A3.

The relationships with regard to the breakage rate k_i are derived on the basis of **OLSEN**'s work [154] with slight modification.

According to **GAUDIN** and **MELOY** [68] the probability of breakage can be given as

$$p_1 = \varkappa_1 d_B d \qquad (A.3.1)$$

for all falling balls, where $\varkappa_1$ is a constant, d_B is the diameter of the ball, d is the diameter (size) of the particle.

In ball mills the distribution of the effective ball energy - according to the opinion of most authors - follows the **BOLTZMANN** distribution, i.e.

$$g(E_n) = \frac{1}{P} \exp \{ \frac{E_n}{P} \} \qquad (A.3.2)$$

where E_n is the effective ball energy, P is the mathematical average of the effective ball energy. The latter is usually called the *mill potential*.

Let P_d be the energy necessary to break a particle of size d. Then the probability of breakage, if the particle collides with the ball, is

$$p_2 = x_2 \int_{P_d}^{\infty} g(E_n)\, dE_n = x_2 \exp\{- \frac{P_d}{P}\} \tag{A.3.3}$$

where (A.3.2) is also taken into consideration. Since the probability of breakages deriving from falling and collision can be regarded as independent, the joint probability of the breakage in the ball mill becomes

$$p = x_1 x_2 d_B\, d \exp\{ - \frac{P_d}{P} \} . \tag{A.3.4}$$

In a short time interval there is a great number of collisions between the particles and balls; therefore the breakage rate is, on the one hand, proportional to the above probability. On the other hand, it is obviously proportional to the average rotation speed of the charge and the number of balls, as well, i.e.

$$k = x_3 d_B n_B\, d\, z \exp\{ - \frac{P_d}{P} \} \tag{A.3.5}$$

where n_B is the number of balls, z is the relative rotation speed of the mill. Since the number of balls is inversely proportional to the third power of the ball diameter for a given charge, k becomes

$$k = x_4 \frac{d\, z}{d_B^2} \exp\{ - \frac{P_d}{P} \} . \tag{A.3.6}$$

We have seen in Fig. 3-6 that the dependence of the breakage rate on the material f to be ground has a strongly nonlinear, extremum character. For small values of f the dependence is directly proportional; for large values of f, however, it is inversely proportional, while at f_0 there is a maximum value. This relationship is represented by the expression

$$k = k_0 \frac{f\, d\, z}{(f_0^2+f^2) d_B^2} \exp\{ - \frac{P_d}{P} \} . \tag{A.3.7}$$

According to experiment, k_0 in (A.3.7) also depends on the size of the particles. Taking this size dependence into consideration by (3.26), we get

$$k = k^* \frac{d^\alpha\, f\, z}{(f_0^2+f^2) d_B^2} \exp\{ - \frac{P_d}{P} \} . \tag{A.3.8}$$

Experiments have shown that P_d is usually size-dependent, while P is not. **OLSEN** [154] simplified the very sophisticated relationships of **LANGEMANN** and he got the equation

$$P \equiv \varkappa_5\, d_B^{3}\, z^2\, (1+\beta_1 z^4+\beta_2 z^8)\ . \qquad (A.3.9)$$

Using this size-dependent expression for P_d, finally we get k_i for the breakage rates

$$k_i = k_i^* \frac{d_i^{\alpha}\, f\, z}{(f_o^2+f^2)\, d_B^2} \exp\left\{ \frac{-\gamma_i}{(d_B^3 z^2 (1+\beta_1 z^4+\beta_2 z^8)} \right\} \qquad (A.3.10)$$

where f_o, β_1 and β_2 are size-independent, while k_i^* and γ_i are size-dependent parameters.

Note that k_i^* can usually be regarded as constant throughout a wide range of small fractions.

Here the variables d_B, f and z can be also taken as control variables.

REFERENCES

1 Adel, G.T., A.G. Ulsoy and K.V.S. Sastry (1980). Dynamic analysis and automatic control of open-circuit grinding mills. Paper submitted for publication in Automatica.

2 Anselm, W. (1950). Zerkleinerungstechnik und Staub, VDI Verlag, Düsseldorf.

3 Apling, A.C., D.H. Montaldo and P.A. Young (1982). The development of a combined comminution - classification flotation model, XIV. Int.Mineral Processing Congress, Toronto,**III**/4.1-4.22.

4 Appiah, R.K. (1976). A dynamic model of a closed-circuit ore grinding system, The University of Zambia, School of Engineering, Department of Electrical Engineering, UNZA/EE/76/02.

5 Appiah, R.K. (1976). Linear dynamical models of a closed-circuit ore grinding system, The University of Zambia, School of Engineering, Department of Electrical Engineering, UNZA/EE/76/03.

6 Arbiter, N. and C. Harris (1965). Particle size distribution - time relationships in comminution. Br. Chem. Eng.,**10**,240-247.

7 Арефьев, Б. А. (1966). К задаче оптимизации процессов в обьектов с замкнутыми циклами. Радиофизика, 9, 817-823.

8 Aström, K.J. (1970). Introduction to stochastic control theory, Academic Press,New York.

9 Aström, K.J. and B. Wittenmark (1973). On self-tuning regulators, Automatica,**9**,185.

10 Auer, A. (1981). Verallgemeinertes lineares Modell des Zerkleinerungsprozesses, Powder Technology,**28**, 65-69.

11 Auer, A. (1981). Die Identifikation des Zerkleinerungsprozesses, Powder Technology,**28**, 71-75.

12 Auer, A. (1981). Das probabilistische Modell des kontinuierlichen Zerkleinern,Powder Technology,**28**, 77-82.

13 Austin, L.G., P.T. Luckie and H.M. Seebach (1976). Optimization of a cement milling circuit with respect to particle size distribution and strength development by simulation models,Dechema Monographien, Band 79,519-537.

14 Austin, L.G. and R.R. Klimpel (1964). The theory of grinding, Industrial and Engineering Chemistry,**56**,19-29.

15 Austin, L.G. (1972). A review. Introduction to the mathematical description of grinding as a rate process, Powder Technology,**5**,1-17.

16 Austin, L.G. and P.T. Luckie (1972). Methods for determination of breakage distribution parameters, Powder Technology,**5**,215-222.

17 Austin, L.G. and V.K. Bhatia (1972). Experimental methods for grinding studies in laboratory mills, Powder Technology,**5,** 261-266.

18 Austin, L.G. and P.T. Luckie (1972). The estimation of non-normalized breakage distribution parameters from batch grinding tests, Powder Technology,**5**,267-271.

19 Austin, L.G., P.T. Luckie and R.R. Klimpel (1972). Solutions of the batch grinding equation leading to ROSIN-RAMMLER distribution, Trans. AIME,**252**,87-94.

20 Austin, L.G., P.T. Luckie and D. Wightman (1975). Steady-state simulation of a cement milling circuit,Int. J. of Mineral Processing,**2**,127-150.

21 Austin, L.G. and P.T. Luckie (1976). Ein empiriches Modell für Windsichterdaten, Zement-Kalk-Gips,452-457.

22 Austin, L.G., K. Shoji and P.T. Luckie (1976). The effect of ball size on mill performance, Powder Technology,**14**,71-79.

23 Austin, L.G., N.P. Weymont and O. Knobloch (1980). The simulation of air-swept cement mills, European Symp. on Particle Technology,640-655.

24 Austin, L.G. and P. Bagga (1981). An analysis of fine dry grinding in ball mills, Powder Technology,**28**,83-90.

25 Balázs, Á., Bojtor J., Veszeli L., Hambulkó J. (1967). Eljárás golyósmalom szabályozására, Magyar szabadalmi leírás, 155756,6.

26 Bang-Pedersen, H. (1975). Complete chemical control of the cement processes, IFAC Congress.

27 Bang-Pedersen, H. and F.L. Smidth (1976). New possibilities in automatic control of cement plants?, IEEE Conference on Cement Industry, Tucson (Arizona,USA),15.

28 Bang-Pedersen, H. (1976). Die Entwicklung der Automation und ihre Anwendung im Zementwerk Gorazdze, Polen,Zement-Kalk-Gips, 162-168.

29 Bányász, Cs. and L. Keviczky (1982). Direct methods for self-tuning PID regulators, 6th IFAC Symp. on Ident. and Syst. Par. Est.,Washington (USA),1249-1254.

30 Barton, A.D., D.G. Jenkins, F.P. Lees and R.W. Murtagh (1970). Commissioning and operation of a direct digital control computer on a cement plant, Measurement and Control,**3**,T77-T88.

31 Bay, T., C.W. Ross, J.C.Andrews and J.L. Gilliland (1968). Dynamic control of the cement process with a digital computer system, IEEE Trans. /IGA,120.

32 Bay, T. and Ch.W. Ross (1975). Multivariable control systems for cement plants, 2nd Annual Advanced Control Conference on Control of Systems with Multiple Interacting Variables, Lafayette (Indiana,USA),7.

33 Beke, B. (1964). Principles of Comminution. Akadémiai Kiadó, Budapest.

34 Beke, B. (1981). The Process of Fine Grinding, Martinus Nijhoff, W. Junk Publ., Boston.

35 Bélanger, P.R. and S.F. Kennedy . Control of a closed circuit cement mill.

36 Bernt, J.O. (1964). Automatic control of feed rates to grinding mills in the cement industry,19th Annual ISA Conference, New York,4.1-3-64,10.

37 Blaine, R.Q. (1949). Tentative method of test for fineness of portland cement by air permeability apparatus, ASTM Standard, 103-109.

38 Блюмкин, Г.В. (1956). Некоторые закономерности и автоматизаций процесса одностадиального замкнутого цикла мокрого измельчения, Ленинградский Горный Институт., 3-15.

39 Bokor, J. and L. Keviczky (1985). Modelling of a blending process by elementary subsystem representation, Int. J. Modelling and Simulation,**5**,3,110-113.

40 Bond, F.C. (1952). The third theory of comminution, Trans. AIME, USA, 484-494.

41 Borisson, U. and R. Syding (1976). Self-tuning control of an ore crusher, Automatica,**12**,1-8.

42 Broadbent, S.R. and T.G. Callcott (1956). A matrix analysis of processes involving particle assemblies, Phil. Trans. R.Soc. Lond.Ser.,A,**249**,99-123.

43 Brown, V.J., V.S. Prozuto and A.A, Trushin . Synthesis of an automatic control system based on a static model of a wet grinding unit.

44 Cassel, H.P. (1972). A control system for a two compartment cement grinding mill,IEEE Cement Industry Technical Conference, Detroit (USA),5.

45 Cooper, A.G. and R.J. Pospisil (1975). A fully automatic rawmix control system, IEEE Cement Industry Technical Conference, Montreal (Canada),11.

46 Costa, H. (1960). Automation of a rotary cement kiln by means of a simple analogue computer.Proc. of 1st IFAC Congress, Moscow,**4**,265-273.

47 Cross, M., R.W. Young, P.E. Wellstead and R.D. Gibson (1977). The mathematical modelling and control aspects of the pelletizing of iron ores, Control Systems Centre, Univ. of Manchester, Inst. of Science and Technology,Manchester,No.374.

48 Csáki F. (1973). Fejezetek a szabályozástechnikából. Állapotegyenletek, Mũszaki Könyvkiadó, Budapest.

49 Csáki F., Kis P., Hilger M., Keviczky L., Kolostori J. (1977). Egy megoldás zárthurku malmok szélsõérték-szabályozására, Mérés és Automatika,**25**,10-13.

50 Csáki F., L. Keviczky, J. Hetthéssy, M. Hilger and J. Kolostori (1978). Simultaneous adaptive control of chemical composition and maximum quantity of ground materials at a closed-circuit ball mill, 7th IFAC Congress, Helsinki,453-460.

51 Domanowski, A., K. Derbsch und H.P. Sommer (1980). Betrachtungen zur Regelbarkeit von Mahl-Sicht-Kreislaufen in der Zementindustrie, Zement-Kalk-Gips,**33**,188-189.

52 Domnrowe, H. und L. Scheer (1973). Einsatz von Mahlhilfsmitteln bei der Zementmahlung, Silikattechnik,**24**, 132-134.

53 Dozortsev, V.M., E.L. Itskovich, I.V. Nikiforov and I.I. Perelman (1983). Computer

control of a cement plant, IFAC/IFIP Symp. on Real Time Digital Control Applications, Guadalajara (Mexico),221-226.

54 Draper, N. and A.J. Lynch (1965). An analysis of the performance of multi-stage grinding and cyclone classification circuits,Proc. Aust. Inst. Min. Metall.,**213**,89-128.

55 Duckworth, G.A. and A.J. Lynch (1982). The effect of some operating variables of autogenous grinding circuits and their implication for control, 14th Int. Mineral Processing Congress, Toronto,**III**/1.1-1.21.

56 Dudnikov, E.E. Optimal Blending of raw materials in cement production.

57 Egger, H., M. Cuenod, P.E. Muller and J.P. Piguet (1971). Some experiences on computer control applications in cement plants.3rd IFAC/IFIP Conf. on Digital Control Appls. to Process Control, Helsinki,**XIV-4**,1-6.

58 Epstein, B. (1948). Logarithmico-normal distributions in breakage of solids, Ind.Eng.Chem.,**40**,2289-2291.

59 Feige, F. (1973). Betrachtungen zur Berechnung des Leistungsbedarfs von Rohrmühlen, Silikattechnik,**24**,58-62.

60 филипенко, В. И. и И. Т. Шепелев (1961). Самонастраивающаяся система авторегулирования узла., Колума, 28-30.

61 Forman, I. und A. Cigánek (1977). Regelung einer Zementumlauf- mahlanlage, Zement-Kalk-Gips,**30**,79-83.

62 Frühwein, P. (1976). Algorithms for estimating the process parameters of continuous grinding, Dechema Monographien,Band 79,505-518.

63 Gardner, R.P. and K. Verghese (1976). Tanks-in-series transient models for the determination of model simulation parameters in continuous, closed-circuit comminution processes, Dechema Monographien, Band 79,489-504.

64 Gardner, R.P. and K. Sukanjnajtee (1972). A combined tracer and back-calculation method for determining particulate breakage functions in ball milling, Part I., Rational and description of the proposed method, Powder and Technology,**6**, 65-74.

65 Gardner, R.P. and K. Sukanjnajtee (1973). A combined tracer and back-calculation method for determining particulate breakage functions in ball milling, Part III., Simulation of an open-circuit continuous milling system, Powder and Technology, **7**,169-179.

66 Gardner, R.P. and K. Verghese (1975). A model with closed form analytic solution for steady-state, closed-circuit comminution processes, Powder technology,**11**,87-88.

67 Gates, A.O. (1913). The crushing surface diagram, Eng. Min.I., **95**,1039-1041.

68 Gaudin, A.M. and T.P. Meloy (1962). Model and comminution distribution equation for single fracture, Trans. AIME,**223**, 40-43.

69 Gautier, E.H.Jr., M.R. Hurlbut and E.A.E. Rich (1971). Recent developments in automation of cement plants, IEEE Trans. Ind.Gen.Appl.,**IGA-7**,459-469.

70 Gel'fand, Ya.Ye., E.L. Itskovich and A.A. Pervozvansky . A hierarchical structure of optimal control algorithms in the chemical industry with special reference to a cement plant.

71 Gerstberger, R. (1970). Statistische Methoden bei der Probenauswertung zur Rohmehlanalyse und deren praktische Anwendung bei der Rohmehlaufbereitung durch Prozessrechner.Zement-Kalk-Gips,**59**,1.

72 Giesa, E. (1971). Über eine zweckmässige Mahltechnologie beim Einsatz von Braunkohlenfilterasche als Zumahlstoff zum Zement, Silikattechnik,**22**,349-352.

73 Godard, P. (1976). La probleme du mélange et son application á la conduite du cru en cimenterie, Automatisme,367-395.

74 Gupta, V.K. and P.C. Kapur. A critical appraisal of the discrete size models of grinding kinetics, Dechema Monographien, Band 79,447-465.

75 Gupta, V.K. and P.C. Kapur (1974). Empirical correlations for the effects of particulate mass and ball size on the selection parameters in the discretized batch grinding equation, Powder Technology,**10**,217-223.

76 Haber, R., I. Vajk and L. Keviczky (1982). Nonlinear system identification by linear systems having signal dependent parameters, 6th IFAC Symp. on Ident. and Syst. Par. Est.,Washington, 421-426.

77 Hammer, H. (1972). Computer controlled rawmeal production in cement industry, Regelungstechnik und Processdaten- verarbeitung,**5**,190-198.

78 Хараш, Л. И. (1961). Самонастраивающаяся система регулирования начальной газопроницаемости агломерационной шихты, Обогащение Руд., **2**,43-47.

79 Harris, C.C. Relationships for the xYt comminution surface.

80 Harris, C.C. (1972). Graphical representation of classifier- corrected performance curves, Trans. Inst. of Mining and Metallurgy,**81**, 243-245.

81 Harris, C.C. and A. Chakravarti (1970). The effect of time in batch grinding, Powder Technology,**4**,57-60.

82 Haskell, P.E. and C.H.J. Beaven . Computer simulation of the transient performance of a closed-circuit grinding system.

83 Hawkins, J. (1971). Using digital measurement in raw mill control systems, IEEE Cement Industry Technical Conference, Washington (USA),9.

84 Heaton, N.M.F. (1971). A linear programming algorithm for on-line cement raw materials blending.3rd IFAC/IFIP Conf. on Digital Control Appls. to Process Control, Helsinki,**XIV-5**,1-8.

85 Heinrich, K.J. und H.G. Minske (1977). Modellvorstellungen und Regelkonzepte für Kugelmühlen in der Zement-industrie, Zement-Kalk-Gips,**30**,273-278.

86 Heiskanen, K. (1976). A method to calculate comminution energies of particle distributions, Dechema-Monographien,Band 79,631-639.

87 Herbst, J.A., G.A. Grandy and D.W. Fuertenau. Population balance models for the design of continuous grinding mills.

88 Herbst, J.A., G.A. Grandy and T.S. Mika (1971). On the development and use of lumped parameter models for continuous open- and closed-circuit grinding systems, Trans. Inst. of Mining and Metallurgy, Sec.C,**80**,C193-C198.

89 Herbst, J.A., T.S. Mika and K.Rajamani (1975). A comparison of distributed and lumped parameter models for open circuit grinding, Dechema-Monographien,Band 79,467-487.

90 Hetthéssy, J., I. Vajk, R. Haber, M. Hilger and L. Keviczky (1983). A microprocessor-based adaptive composition control system, Proc. of IFAC/IFIP Symp. on Real-Time Digital Control Appls., Guadalajara (Mexico),209-213.

91 Higham, J.D. (1971). Dynamic computer regulation of a dry process cement kiln, Proc. of IEE,**118**,3-4,609-617.

92 Hilger, M., R. Haber and L. Keviczky (1975). DDC of raw material blending, simulation investigations, United Kingdom Simulation Conference, Bowness-on-Windermere.

93 Hilger, M., Kolostori J., Keviczly L. (1978). Az adaptiv szabályozással kapcsolatos vizsgálatok a nyerslisztstabilizálásnál és a körfolyamatos õrlésnél, Épitõanyag, **30**,2,48-53.

94 Hilger, M., J. Kolostori und L. Keviczky (1978). Untersuchungen über die adaptive Regelung bei der Rohmehlstabilisierung und der Kreislaufmahlung, Zement-Kalk-Gips,**31**,175-178.

95 Hilger, M., L. Keviczky, J. Kolostori und F. Szijj (1978). Moderne Regelungsmethoden und die bei ihrer Anwendung erworbenen Erfahrungen, 2. Fachtagung "Regelungstechnik in Zementwerken, Bielefeld, 1-24.

96 Hilger, M., L. Keviczky und J. Kolostori (1979). Einige Fragen zur Modellisierung und Mengenregelung der Kreislaufmühlen in der Zementindustrie, Zement-Kalk-Gips,**32**,499-503.

97 Hilger, M., J. Hetthéssy, L. Keviczky und J. Kolostori (1981). Vergleichende Untersuchungen zur Regelung von im Kreislauf geschalteten Kugelmühlen aufgrund von Betriebsergebnissen, Zement-Kalk-Gips,**34**,146-150.

98 Hilger, M., L. Keviczky und J. Kolostori (1981). Möglichkeiten der Verminderung des spezifischen Energieverbrauchs von Umlaufmühlen, Preprints Fachtagung "Zerkleinerung und Tribochemie".

99 Hilger, M., L. Keviczky und J. Kolostori (1982). Systemtechnische Beschreibung von Umlauf-Mahlprozessen in der Zementindustrie, 8. Internationale Baustoff- und Silicattagung, Weimar,**2**,165-169.

100 Hodouin, D., J. McMullen and M.D. Everell (1980). Mathematical simulation of the operation of a three-stage grinding circuit for a fine grained Zn/Pb/Co ore, European Symp. Particle Technology,686-702.

101 Hoenig, H. (1972). Component control in a cement plant using a process computer, Zement-Kalk-Gips,**1**,31-36.

102 Horst, W.E. and E.J. Freeh (1970). Mathematical modelling applied to analysis and control of grinding circuits, AIME Ann. Meeting, Denver,70-B-27.

103 Hubbard, M. and T.Da Silva (1982). Estimation of feedstream concentrations in cement raw material blending, Automatica,**18**,595-606.

104 Hukki, R.T. About the ways and means to improve the performance of the closed grinding circuit.

105 Hürliman, W. (1962). Steuerung und Regelung neuzeitlicher Mahlanlagen, Zement-Kalk-Gips,383-390.

106 Inove, T., K. Okaya and T. Imaizumi (1980). A dynamic simulation model for the closed circuit grinding process,European Symp. Particle Technology,656-671.

107 Jaspan, R.K., H.W. Kropholler, T. Mika and E. Woodburn (1976). An analysis of closed circuit wet-grinding mill control characteristics by simulation, Dechema-Monographien,Band 79, 539-557.

108 Kaiser, V.A. (1970). Computer control in the cement industry, Proc. IEEE,**58**,70-77.

109 Kapur, P.C. and P.K. Agrawal (1970). Effect of feed charge weight on the rate of breakage of particles in batch grinding, Trans. Inst. of Mining and Metallurgy, Sec.C,C269-C272.

110 Karra, V.K. and D.W. Fuerstenau (1977). The scale-up of grate-discharge continuous ball mills, Int. J. of Mineral Processing,**4**,1-6.

111 Karra, V.K., S.H.R. Swaroop and D.W. Fuerstenau (1980). Grinding in grate-discharge ball mills: the effect of the mill length, European Symp. Particle Technology,Amsterdam,493-505.

112 Kelsall, D.F. and K.J. Reid (1965). The derivation of a mathematical model for breakage in a small, continuous, wet ball mill, AIChE Symp. Series,**4**,14-20.

113 Kelsall, D.F., P.S.B. Stewart and K.J. Reid (1968). Confirmation of a dynamic model of closed-circuit grinding with a wet ball mill, Trans. Inst. of Mining and Metallurgy, Sec.C,**77**,C120-C127.

114 Kelly, M. (1971). The effect of time in batch grinding, Powder Technology,**4**,56-60.

115 Keviczky, L., Cs. Bányász, M. Hilger and F. Hamikus (1975). A new simulation model of ball grinding mills for control purposes, Summer Simulation Conference,San Francisco (USA),6.

116 Keviczky, L., J. Hetthéssy, M. Hilger and J. Kolostori (1976). Self-tuning computer control of cement raw material blending, IFAC/IFIP SOCOCO Conference, Tallin (USSR),9-14.

117 Keviczky, L., I. Udvardi and Cs. Bányász (1976). Optimization of a cement mill by correlation technique - Simulation results, 8th AICA Congress,Delft,737-745.

118 Keviczky, L. (1976). Nonlinear dynamic identification of cement mill to be optimized, 4th IFAC Symp. on Ident. and Syst. Par. Est., Tbilisi,388-396.

119 Keviczky, L., J. Hetthéssy, M. Hilger and J. Kolostori (1978). Self-tuning adaptive control of cement raw material blending, Automatica,**14**,525-532.

120 Keviczky, L., Cs. Bányász, M. Hilger and J. Kolostori (1979). CAD of closed circuit mills, IFAC CAD Congress, Zürich, 497-503.

121 Keviczky, L., I. Vajk and J. Hetthéssy (1979). A self-tuning extremal controller for the generalized Hammerstein model, 5th IFAC Symp. on Ident. and Syst. Par. Est.,Darmstadt,1147-1151.

122 Keviczky, L., M. Hilger and J. Kolostori (1982). On control engineering models of cement grinding mills, 14th Int.Congress on Mineral Processing,Toronto,**III**/12.1-12.12.

123 Keviczky, L. (1983). Control in cement production (invited plenary paper), 4th IFAC Symp. on Automation in Mining, Mineral and Metal Processing,Helsinki.

124 Kihlstedt, P.G. (1962). The relationship between particle size distribution and specific surface in comminution, Symp. Zerkleinern Verlag Chemie,Weinheim VDI Verlag, Düsseldorf, 205-216.

125 Kihlstedt, P.G. (1964). Assessment of comminution by means of particle size and specific surface, 7th Int.Congress on Mineral Processing,11-17.

126 Klee, B.J. (1976). Why particle size measurement is key to increased grinding tonnage?, World Mining,30-35.

127 Kolostori, J., R. Haber, L. Keviczky and M. Hilger (1976). On simultaneous optimal control of raw material blending and a ball grinding mill, 2nd IFAC Symp. on Automation in Mining, Mineral and Metal Processing,Johannesburg,143-158.

128 Kolostori, J.,Hilger M., Keviczky L. (1977). Cementipari nyersanyagelõkészítés és a cementõrlés energetikai kérdései, SILICONF, Budapest.

129 Kolostori, J., Pethõ Sz., Hilger M. (1979). Korszer~u cementipari technológiák, BME TKI,Budapest.

130 Koulen, K. und H. Schneider (1965). Zur Berechnung des Gewichtsausbringens bei der Sichtung, Auchbereitungs Tech., **6**, 586-589.

131 Kunze, E. (1975). Einige Eigenschaften direkt adaptierender Regelungsverfahren am Beispiel einer Kugelmühle, ITTB- Mitteilungen,36-40.

132 Kunze, E. und M. Salaba (1979). Praktische Erprobung eines adaptiven Regelungsverfahrens an einer Zementmahlanlage,PDV- Bericht Kfk-PDV 158, Kernforschungszentrum, Karlsruhe GmbH, 1-66.

133 LeHouillier, R., A.Van Neste and J.C. Marchand (1977). Influence of charge on the parameters of the batch grinding equation and its implications in simulation, Powder Technology,**16**,7-15.

134 Lenning, R.L. (1972). Adaptive control for a grinding machine, United States Patent,3,699.720,7.

135 Lin, C.S. and N.F. Gardner (1974). Application of disturbance observer to computer control of blending process. JACC,Austin (Texas),514-519.

136 Loveday, B.K. (1967). An analysis of comminution kinetics in terms of size distribution parameters, J. of South African Inst. Min. Metall.,**68**,111-131.

137 Luckie, P.T. and L.G. Austin (1973). Technique for derivation of selectivity functions from experimental data.10th Mineral Process Congress,**12**,1-18.

138 Luckie, P.T. (1976). The hybrid model for steady-state mill simulation, Powder Technology,**13**,289-290.

139 Lundan, A. and O. Mattila (1974). A system for the control of the homogenisation of the cement raw meal, 4th IFAC Symp. on Digital Appls. to Process Control, Zürich,**1**,,424-435.

140 Lynch, A.J., W.J. Whiten and N. Draper (1969). Developing the optimum performance of a multistage grinding circuit, Trans. Inst. of Mining and Metallurgy,**79**,169-182.

141 Lynch, A.J. (1977). Mineral crushing and grinding circuits. Their simulation, optimisation, design and control, Elsevier Scientific Publ. Co.,Amsterdam,342.

142 Magallon, R. and J. Santodomingo (1973). Optimizaciøn del Ilenado de molino, Cementi-Mormigon,719-727.

143 Malghan, S.G. and D.W. Fuerstenau (1976). The scale-up of ball mills using population balance models and specific power input, Dechema-Monographien, Band 79,613-630.

144 Meloy, T.P. and B.H. Bergström. Matrix simulation of ball mill circuits considering impact and attrition grinding, 7th Int. Congress on Mineral Processing,19-31.

145 Molerus, O. und H. Hoffmann (1969). Darstellung von Windsichtertrennkurven durch ein stochastisches Modell, Chemie Ingr. Tech.,**41**,340-344.

146 Moreira, R.M., J.O.N.M. De Castro and R.P.Gardner (1972). A combined tracer and back-calculation method for determining particulate breakage functions in ball milling, Part II. Application to hemitate iron ore in a batch laboratory mill, Powder Technology,75-83.

147 Mular, A.L. (1970). The determination of selection elements and lumped parameters for grinding mills, CIM Bulletin, 821-826.

148 Mular, A.L. (1971). Mathematical models for optimum design of grinding circuits, CIM Bulletin,341.

149 Mular, A.L, R.G. Brandburn, B.C. Flintoff and C.R. Larsen (1976). Mass balance of a grinding circuit, CIM Bulletin, 124-129.

150 Muller, T.B. and T.L. Johnson. Simulation of mass, heat and particulate balances in the Bayer precipitation.

151 Naevdal, S., T.Wigen and A.B. Aune . Stockpile prehomogenisation and rawmeal plant with aerofall mill.

152 Nepomnyashchy, E.A. (1980). The fundamentals of comminution theory and calculations of kinetic dispersity, European Symp. Particle Technology,Amsterdam,255-264.

153 Ohta, T. and K. Ishida (1983). Real-time digital control systems for the cement industry, Proc. of IFAC/IFIP Symp. on Real-Time Digital Control Appls.,Guadalajara (Mexico),197-202.

154 Olsen, T. (1972). Modelling and control of ball mill grinding, Universitetet i Trondheim, Norges Tekniske Hogskole, Inst. for Reguleringteknik,72-86-W.

155 Olsen, T. (1973). Modelling and optimization of closed-circuit ball mill grinding, IFAC Symp. on Automated Control in Mining, Mineral and Metal Processing.

156 Olsen, T., H. Besstad and S. Danielsen (1973). Automatic control of continuous autogeneous grinding, IFAC Symp. on Automated Control in Mining, Mineral and Metal Processing.

157 Olsen, T. and S.R. Krogh (1973). Mathematical model of grinding at different conditions in ball mills, AIME Trans, **254.**

158 Olsen, T. (1974). Modelling continuous wet ball mill grinding, IFAC Symp. on Multivariable Technological Systems, Manchester,S18.1-S18.4.

159 Olsen, T. and S.R. Krogh (1975). A low order model of continuous ball mill grinding, 11th Int. Mineral Congress, Cagliari (Italy).

160 Onuma, E., N. Asai and G. Jimto (1976). Analysis of the operating characteristics of steady-state closed-circuit ball mill grinding,Dechema-Monographien,Band 79,559-573.

161 Otomo, T., T. Nakagawa and H. Akaike (1972). Statistical approach to computer control of cement rotary kilns, Automatica,**8**,35-48.

162 Pennel, A.R. (1976). The application and performance of a predictive controller for cement raw meal blending systems, IEEE Cement Industry Technical Conference,Tucson (Arizona),30.

163 Phillips, R.A. Automation of a portland cement plant using a digital control computer.

164 Plitt, L.R. (1971). The analysis of solid-solid separations in classifiers, CIM Bulletin,**64**,42-47.

165 Plitt, L.R. (1976). A mathematical model of the hydrocyclone classifier, CIM Bulletin, 114-123.

166 Porter, E.S. and J. Warshawsky (1969). Automatic sampling and measurement of surface area of pulverized material, IEEE Trans. Ind. Gen. Appl.IGA-5,773-778.

167 Ragot, J., M.Roesch, P. Degoul and V. Berube (1976). Transient study of a closed grinding circuit, Proceedings 2nd IFAC Symp. on Automation in Mining, Mineral and Metal Processing, Johannesburg,129-142.

168 Rajamani, K. and J.A. Herbst (1980). A dynamic simulator for the evaluation of grinding circuit control strategies, European Symp. Particle Technology,Amsterdam,64-81.

169 Rammler, E. und A. Bahr (1971). Zur Ermittlung der spezifischen Oberfläche von Korngrössen-Verteilungen, Teil I: Allgemein anwendbare Verfahren, Verfahrenstechnik,483-491.

170 Rammler, E. und A. Bahr (1972). Zur Ermittlung der spezifischen Oberfläche von Korngrössen-Verteilungen, Teil II: Rechnerische Verfahren für spezielle Korngrössenverteilungen, Verfahrenstechnik,**6**,312-320.

171 Rammler, E. und A. Bahr (1972). Korngrössenverteilungen, Teil I: Vergleich von Korngrössenverteilungen, Chem.Techn.,**24**, 345-351.

172 Rammler, E. und A. Bahr (1972). Korngrössenverteilungen, Teil II: Vergleich von Körnungsnetzen, Chem.Techn.,**24**,738-743.

173 Randolph, A.D. and R. Ranjan (1977). Effect of a material-flow model in prediction of particle size distributions in open- and closed-circuit mills, Int. J. Mineral Processing,**4**, 99-110.

174 Rao, P.R. and R. Sankaran. Identification of a batch grinding mill.497-503.

175 Reid, K.J. (1965). A solution to the batch grinding equation, Chem. Eng. Sci,**20**,953-963.

176 Reid, K.J. (1971). Derivation of an equation for classifier-reduced performance curves, Can. Metall.Q.,**10**, 253-254.

177 Reig, C. M. (1972). Einflussfaktoren auf die Verweilzeit in Zementmühlen, Zement-Kalk-Gips,**25**,245-247.

178 Rich, E.A.E. (1965). Cement automation, IEEE Cement Industry Technical Conference, Allentown (Pennsylvania),17.

179 Riegel, R.W. Improved process control of raw material blending and kiln/cooler operation result in major energy savings.

180 Riegel, R.W. (1964). Process analysis - an integral part of automation, IEEE Cement Industry Conference,Pasadena (California),105-111.

181 Rittinger, R.P. (1867). Textbook of Mineral Dressing, Ernst and Korn, Berlin.

182 Rodd, M.G. and J.H. Potgieter (1976). A microprocessor based weighing and feed control system, Proc. 2nd IFAC Symp. on Automation in Mining, Mineral and Metal Processing, Johannesburg,381-388.

183 Roger, R.S.C, A.M. Shoji, A.M. Hukki and R.J. Linn (1982). The effect of liner design on the performance of a continuous wet ball mill, 14th Int. Mineral Processing Congress, Toronto,**I**/5.1-5.20.

184 Romig, J.R., W.R. Morton and R.A. Philips (1965). Making cement with a computer control system, IEEE Cement Industry Technical Conference, Allentown (Pennsylvania),13.

185 Rose, H.E.(1957). A mathematical analysis of the internal dynamics of the ball mill on the basis of probability theory, Trans. Inst.Chem. Engr.,**35**,87-97.

186 Rose, H.E. (1975). On the comminution process in a ball mill, Dechema-Monographien, Band 79,253-266.

187 Rosin, P. and E. Rammler (1933). The laws governing the fineness of powdered coal, J. Inst. Fuel,**7**,29-36.

188 Ross, Ch.W. and T. Bay (1977).Energy conservation through automatic controls in the

cement industry, 4th Annual Advanced Control Conference on Conservation of Process Energy, Chicago (Illinois),1-9.

189 Rossberg, R. Mischungsregelung für Zement-Rohmehl- Aufbereitung,Einzelbericht 0273531 zu H und Messwerk.

190 Sastry, K.V.S. and J.S. Wakeman (1980). Design and control aspects of regrinding mill circuits, European Symp. Particle Technology,672-685.

191 Schaknies, G. (1969). Zur Optimierung einer Füllstandsregelung bei Kugelmühlen, Zement-Kalk-Gips,**19**,24-31.

192 Schaknies, G. (1969).Versuche zum Verhalten eines Mühlenumlaufsystems, Zement-Kalk-Gips,**19**,212-217.

193 Schaknies, G. (1975). Parameterarme Prozessmodelle, ein Hilfsmittel für die praktische Prozessführung, Dechema- Monographien,Band 79,599-612.

194 Schaknies, G. (1977). Signalanalyse, ein Mittel zur optimalen Prozessführung, Zement-Kalk-Gips,**30**,71-74.

195 Schink, H. (1962). Überwachung und Regelung von Zerkleinerungsmaschinen, Zement-Kalk-Gips,**12**,409-411.

196 Schramm, R. (1973). Vorschlag für ein einfaches Mahlanlagenmodell, Silikattechnik,**24**,124-126.

197 Schramm, R. und E. Gaitsch (1974). Eine quantitative Methode zur Gattierung in Kugalmühlen, Zement-Kalk-Gips,**27**,330-332.

198 Schubert, H. (1964). Aufbereitung fester mineralischer Rohstoffe, Band 1, Veb Deutcher Verlag für Grundstoff- industrie, Leipzig.

199 Schulz, R. (1973). Die Regelung unter Einwirkung Stochastischer Störungen, Messen-Prüfen,266-270.

200 Schulz, R. (1973). Adaptive Regelung einer Rohmehlmahlanlage, Sprechsaal,**106**,223-225.

201 Schulz, R. (1977). Leistungsregelung einer Kugelmühle, Zement-Kalk-Gips,**30**,49-52.

202 Schulze, H. (1970). Regelung und Optimierung von Kugelmühlen, Siemens Aktiengesellschaft. Automatisierungstechnik Zement. Software-Information,15.

203 Schulze, H. (1971). Anwendung von Schätzverfahren für die Korngrössen von Regelstrecken aufgrund von Messungen im geschlossenen Regelkreis, Regelungstech. und Prozess-Datenverarb.,**19**,113-119.

204 Schulze, H. (1974). Adaptive Verfahren zur digitalen Regelung von Kugelmühlen in Zementwerken, Regelungstech. und Prozess-Datenverarb.,**22**,174-177.

205 Schulz, R. (1983). Adaptive control of a ball mill with self-tuning reference model, Proc. of IFAC/IFIP Symp. on Real-Time Digital Control Appls.,Guadalajara (Mexico),203-207.

206 Shoji, K., S. Lohrasb and L.G. Austin (1980). The variation of breakage parameters with ball and powder loading in dry ball milling, Powder Technology,**25**,109-114.

207 Simmons, R.I (1968). Complete cement plant automation, Measurement and Control (GB),B1,T94-T95.

208 Soltynski, A. and S. Pampuch (1975). Automatyzacja procesu zestawiania mieszaniny surowców w cementowni, Cement-Wapno- Gips,31-37.

209 Штейн, С. А., А.А. Ферштенфельд, Э. Ф. Меднис и Е. Л. Крицкий (1957). Сравнительные ичпытания различных способов автоматического регулирования шаровых мельниц, Обогащение Руд., 55-66.

210 Surmann, W. (1962). Rohmühlen und Mahltechnik,Zement-Kalk- Gips,**12**,89-97.

211 Talabér, J., M. Hilger, F. Csáki and L. Keviczky (1977). Modern control concepts in the cement industry, 5th IFAC/IFIP Conference on Digital Computer Applications to Process Control, Hague,383-393.

212 Tamura, K. and T. Tanaka (1970). Rate of ball milling and vibration milling on the basis of the comminution law. Probability Theorem, Ind. Eng. Chem. Process Des. Develop.,**9**, 165-173.

213 Tarján, G. (1969). Ásványelőkészítés I.,Tankönyvkiadó, Budapest.

214 Teutenberg, J. (1971). Prozessautomation einer Zementfabrik in Argentinien, Zement-Kalk-Gips,**4**,141-151.

215 Teutenberg, J. (1973). Laboratory automation in cement works, Cement Technology,131-140.

216 Teutenberg, J. (1981). Dezentrale Prozessautomatisierung in der Zementindustrie, Zement-Kalk-Gips,**34**,113-122.

217 Tiggesbäeumker, P. and J. Williams (1976). Large mills for dry raw material and clinker grinding, IEEE Trans. on Ind. Appl., **IA-12**,104-119.

218 Тихонов, О. Н. (1961). Самонастраивающаяся система регулирования измельчительного с замкнутум циклом. Обогащение Руд., 44-49.

219 Тихонов, О. Н. (1963). Автоматизация процесса измельчения на обогатительных фабриках, Государственный комитет по чёрной и цветной металлургии при ГОСПЛАНЕ СССР, Москва, 3-51.

220 Trawinski, H. (1976). Die mathematische Formulierung der Tromp-Kurve, Teil I., Aufbereitungs-Technik,248-254.

221 Trawinski, H. (1976). Die mathematische Formulierung der Tromp-Kurve, Teil II., Aufbereitungs-Technik,449-459.

222 Tromp, K.F. (1937). Neue Wege für die Beurteilung der Aufbereitung von Steinkohlen, Glückauf,**73**,125-131.

223 Vaillant, A. (1971). Classifier selectivity functions with two and three parameters, Unpublished work.

224 Вальшонок, А. М., И. Б. Финкельштейн, Э. Я. Гутерман, Я.Е. Гельфанд и Ю. И. Цира (1972). Устроуство для автоматического управления процессом измельчения материала в шаровой мельнице. ВИАСМ Изобретение, М.Кл. В02С 25/00.

225 Watson, D., R.W.G. Cropton and G.F. Brockes (1970). Modelling methods for a grinding classification circuit and the problem of plant control.

226 Wellstead, P.E., M.Cross, N. Munro and D. Ibrahim (1977). On the design and assessment of control schemes for balling drum circuits used in pelletizing, Control Systems Centre, Univ. of Manchester, Inst. of Sci. and Technology,Manchester.

227 Wellstead, P.E. and N. Munro (1977). Multivariable control of a cold iron ore agglomeration plant, Control Systems Centre, Univ. of Manchester, Inst. of Sci. and Technology,Manchester, Report 377.

228 Westerlund, T., H. Toivonen and K.E. Nyman (1978). Stochastic and self-tuning control of continuous cement raw material mixing system, Abo Akademi, The Institution for Automatic Control, Finland.

229 Westerlund, T., H. Toivonen and K.E. Nyman(1980). Stochastic modelling and self-tuning control of a continuous cement raw material mixing system, Modelling, Identification and Control,**1**,17-37.

230 Westerlund, T. (1981). A digital quality control system for an industrial dry process rotary cement kiln, IEEE Trans. on Aut. Control,**AC-26**,4,885-890.

231 Westerlund, T. (1983). Experiences from a digital quality control system for cement kilns, Proc. of IFAC/IFIP Symp. on Real-Time Digital Control Appls.,Guadalajara (Mexico),215-219.

232 Wieland, W. (1967). Automatic fineness control, Minerals Processing, 30-31.

233 Will, E. (1975). Mischungssteuerung mit einem Kleinrechner im Zementwerk Göllheim, Zement-Kalk-Gips,**28**,146-150.

234 Williams, J.C. and M.A. Rahman (1972). Prediction of the performance of continuous mixers for particulate solids using residence time distributions. Part II. Experimental, Powder Technology,307-316.

235 Willis, V. (1968). The symposium on cement automation, Measurement and Control (GB),**1**,T85-T86.

236 Willis, V. and R.T. Simmons (1971). The effect of the computer on the design, control and management of cement processes.3rd IFAC/IFIP Conf. on Digital Computer Appls. to Process Control,Helsinki,**XIV-3**,1-9.

237 Wilson, J. and H.C. Iten (1975). Optimizing energy utilization in cement plant operations.

238 Wolff, G. (1975). Rechnerautomation für ein Wärmetauscher-Drehofen-Kühlersystem mit einem Prozessmodell, Zement-Kalk-Gips,**28**,140-145.

239 Young, S.C.K. (1968). Computer control of raw material blending for the cement industry, Measurement and Control (GB),**1**,T87-T90.

240 Young, S.C.K. (1971). Raw material blending - a multivariable control problem, 4th UKAC Control Convention on Multivariable Control System Design and Appls.,Manchester, 76-80.

241 Zhivoglyadov, V.P. (1970). Optimization of closed-loop systems of control with adaptive distributive control of indirect indices. Avtomatika i Telemechanika,**10**,83-90.

242 Zhivoglyadov, V.P. and B.M. Mirkin (1974). Adaptation and direct digital control of the cement industry technological processes with application of distributed check.4th IFAC/IFIP Conf. on Digital Computer Appls. to Process Control,Zurich,**1**, 451-462.

243 Zins, R. (1973). Stand und Entwicklung getriebloser Rohrmühlenantriebe, Zement-Kalk-Gips,**26**,579-582.

INDEX